Additive Manufacturing of High Temperature Alloys

Additive Manufacturing of High Temperature Alloys

Guest Editors

Vera Popovich
Ehsan Hosseini

MDPI

Basel • Beijing • Wuhan • Barcelona • Belgrade • Novi Sad • Cluj • Manchester

Guest Editors
Vera Popovich
Materials Science and Engineering
TUDelft, Delft University of Technology
Delft
The Netherlands

Ehsan Hosseini
Experimental Continuum Mechanics
EMPA, Swiss Federal Laboratories for Materials Science and Technology
Dübendorf
Swidzerland

Editorial Office
MDPI AG
Grosspeteranlage 5
4052 Basel, Switzerland

This is a reprint of the Special Issue, published open access by the journal *Metals* (ISSN 2075-4701), freely accessible at: https://www.mdpi.com/journal/metals/special_issues/AM_High_Temperature_Alloys.

For citation purposes, cite each article independently as indicated on the article page online and as indicated below:

Lastname, A.A.; Lastname, B.B. Article Title. *Journal Name* **Year**, *Volume Number*, Page Range.

ISBN 978-3-7258-4287-2 (Hbk)
ISBN 978-3-7258-4288-9 (PDF)
https://doi.org/10.3390/books978-3-7258-4288-9

Contents

metals

MDPI

Article

Laser Preheating for Hot Crack Reduction in Direct Metal Deposition of Inconel 738LC

Fabian Soffel [1,*], Konrad Papis [1], Markus Bambach [2] and Konrad Wegener [3]

1 Inspire AG, Technoparkstrasse 1, 8005 Zürich, Switzerland; papis@inspire.ethz.ch
2 Advanced Manufacturing Laboratory, ETH Zürich, Technoparkstrasse 1, 8005 Zürich, Switzerland; mbambach@ethz.ch
3 Institute of Machine Tools and Manufacturing, ETH Zürich, Leonhardstrasse 21, 8092 Zürich, Switzerland; wegener@iwf.mavt.ethz.ch
* Correspondence: soffel@inspire.ethz.ch; Tel.: +41-44-633-0849; Fax: +41-44-633-1492

Abstract: Welding of precipitation-hardenable nickel-based super alloys that contain large amounts of Al and Ti is challenging due to their high susceptibility to hot cracking. For metal additive manufacturing (AM) by powder bed fusion (PBF) or direct metal deposition (DMD), various welding process adjustments may prevent the formation of cracks. The aim of this study is the development and experimental characterization of a laser preheating process for DMD of Inconel 738LC. Metallographic cross-sections of multiple test specimens were analyzed to quantify the effect of initial substrate temperature, specimen geometry, deposition parameters, and scanning strategy on the resulting crack density. The results show that increased substrate temperature by laser preheating and reduced specimen size leads to crack-free deposited structures. Therefore, the proposed preheating process may be applied for part fabrication or repair by DMD to reduce or even completely prevent the risk of hot cracking.

Keywords: additive manufacturing; direct metal deposition; preheating; isothermal welding; super alloy; Inconel; hot cracking

1. Introduction

Metal additive manufacturing (AM) typically refers to processes based on welding, such as laser powder bed fusion (PBF-L), electron beam powder bed fusion (PBF-EB), and direct metal deposition (DMD). Powder bed fusion describes a two-step process principle whereby, first, a mechanical system distributes a powder layer and, second, a heat source locally melts the material [1–4]. Direct metal deposition is a single-step process in which carrier gas continuously transports powder into the melt pool [5–7]. This characteristic of DMD is especially advantageous for repair applications or the fabrication of larger parts because it enables the deposition on arbitrary-shaped structures [8]. Precipitation-strengthened nickel-based super alloys are widely used in the energy and aerospace industries due to their excellent corrosion and high-temperature properties [9,10]. However, large amounts of Al and Ti that are added to generate $Ni_3(Al,Ti)$ γ'-precipitates lead to poor weldability [11,12], and hot crack formation significantly limits AM of these alloys. The high susceptibility to hot cracking of this alloy system was shown, for example, for Inconel 738LC, CM247LC, K417G, and Waspaloy [13–16]. DebRoy et al. [17] stated the need to deeper understand various hot cracking phenomena, and several previous studies focused on their qualitative and quantitative description.

The main mechanisms for crack formation in nickel-based super alloys during welding and metal AM processes are solidification and liquation cracking. According to Kou [18], solidification cracks occur in the fusion zone during the terminal stage of solidification, while liquation cracks develop in the partially melted zone due to grain boundary liquation. Xu et al. [19] investigated crack formation in DMD of Inconel 738 and mentioned that the control of crack formation requires a clarification of its mechanism. However, Chauvet et al. [20] remarked that the distinction between the two main crack mechanisms is not straightforward in practice because both require the presence of liquid films, and the intrinsic heat treatment during metal AM may change the crack morphology. Rickenbacher et al. [13] processed Inconel 738LC by PBF-L and found intergranular microcracks with lengths in the range of 50 µm to 250 µm. Adegoke et al. [14] found comparable crack lengths for PBF-L of alloy 247LC. Grange et al. [21] investigated the microcrack surface for Inconel 738LC processed by PBF-L more in detail by bending a test specimen up to failure and analyzing the crack surface by scanning electron microscopy (SEM). They observed a regularly oriented dendritic microstructure and reported lack of liquid metal feeding in the terminal solidification stage as an indicator of solidification cracking.

For direct metal deposition of γ'-strengthened nickel-based super alloys, several studies detected crack propagation over multiple deposited layers that led to significantly larger crack lengths. Liu et al. [16] used DMD for the repair of Waspaloy components and found crack propagation and liquation along grain boundaries. Liu et al. [15] observed multiple crack mechanisms for DMD with alloy K417G and identified liquation cracking as the most common type due to the formation of low-melting eutectics along the grain boundaries. When re-heated above the eutectic temperature during deposition of the following layers, these eutectics may form liquid films that, in combination with contraction, lead to intergranular cracks with a high potential for crack propagation over multiple deposited layers. Without process adjustments, Seidel et al. [22] found crack propagation throughout the entire height of a multilayer structure deposited by DMD with a γ'-strengthened nickel-based super alloy. Therefore, adjustments of the AM process seem to be mandatory to avoid crack formation.

The prevention of hot cracking for this alloy system is possible by several adjustments before and during the welding process. For example, alloy modifications and substrate preheating are upstream measures, while induction-assisted welding and deposition parameter variations are applied during the process. Engeli et al. [23] processed different batches of Inconel 738LC by PBF-L and found that reduced contents of Si lead to lower crack densities. Based on this finding, the same author [24] proposed a modified alloy with reduced contents of Si and Zr and showed a reduction of crack density by a factor of 6. Similarly, Griffiths et al. [25] investigated a modified alloy CM247LC with a reduced amount of Hf for two PBF-L process conditions and, compared to a non-modified alloy, the crack density decreased by 36% and 75%, respectively. Chiang and Chen [26] used laser deposition welding with Inconel 738 powder and inductive preheating at approximately 800 °C to obtain crack-free butt joints. For welding of γ'-strengthened nickel-based super alloys, Stueber et al. [27] proposed to maintain the weld area and its adjacent region at a ductile temperature during the solidification of the weld in the range of between 760 °C and 1093 °C, which is in between the aging and incipient melting temperature of the alloy. Such an isothermal welding process was realized by Xu et al. [28] for Inconel 738LC by inductive substrate heating up to 1050 °C and DMD, which led to crack-free deposited structures. Similarly, Seidel et al. [22] obtained crack-free CM247LC structures by induction-assisted DMD. Several studies focused on the variation of deposition parameters. Cloots et al. [29] varied the scanning speed in PBF-L of Inconel 738LC and found a significantly reduced crack density for higher speed levels. In contrast, Ramakrishnan and Dinda [30] found a reduced cracking susceptibility for lower scanning speeds for DMD of the same alloy. Lee et al. [31] investigated different scanning strategies for PBF-EB of Inconel 738LC to fabricate airfoil structures and found reduced crack formation and lower distortion for modified scan patterns that presumably led to more uniform stress distributions. Based on the findings of previous studies, preheating and isothermal welding seem to have the largest potential to fabricate crack-free structures by DMD of Inconel 738LC. However, as described by Cortina et al. [32], many modern DMD systems are combined machine tools with five-axis configurations. Due to these configurations, the integration of an additional heating system as demonstrated by Seidel et al. [22] and Xu et al. [28] may be technically challenging and not economically feasible, such that novel concepts should be evaluated. Therefore, preheating by the laser of the DMD system in combination with additional process optimizations might be an alternative solution for the crack-free deposition of γ'-strengthened nickel-based super alloys within such machine tools.

The aim of this study is to develop a laser preheating process and to investigate its potential for direct metal deposition of Inconel 738LC in combination with other process adjustments. These adjustments include a variation of the specimen geometry, deposition parameters, and scanning strategy. In this paper, Section 2 describes the experimental procedure; Section 3 presents and discusses the results of the quantitative and qualitative crack analysis; and, finally, Section 4 summarizes the main conclusions.

2. Materials and Methods

A five-axis GF HPM 450 U (Georg Fischer AG, Schaffhausen, Switzerland) direct metal deposition (DMD) and milling machine with an Ambit S5 laser processing system from HMT (Hybrid Manufacturing Technologies, Midlands, UK) was used for part fabrication. The laser system included an IPG YLR-1000-MM-WC (IPG Photonics Corporation, Oxford, MA, USA) fiber laser with a wave length of 1070 nm and a water-cooled DMD processing head with coaxial powder feeding. The working distance of the processing head was set at 9.0 mm, which resulted in an approximate melt pool width of 2.2 mm. Argon was selected as local shielding, powder carrier, and nozzle protection gas and supplied at flow rates of 8 L/min, 6 L/min, and 4 L/min, respectively. Table 1 lists the chemical composition of the Inconel 738LC metal powder. The substrates, on which the specimens were deposited, were 1.4404 stainless steel plates with a length of 50.0 mm, a width of 25.0 mm, and a thickness of 8.0 mm.

Table 1. Chemical composition of the Inconel 738LC metal powder.

Element	Al	C	Co	Cr	Mo	Nb	Ni	Si	Ta	Ti	W	Zr	Others
wt.%	3.44	0.10	8.54	16.10	1.72	0.83	Bal.	0.09	1.73	3.41	2.55	0.06	<0.19

Figure 1 shows the specimen geometries and the investigated scanning vector orientations. Cuboids with edge lengths a = 5 mm, a = 10 mm, and a = 15 mm were chosen as sample geometry to investigate the effect of part size on crack formation. They were deposited with a 0°/90° bidirectional raster scanning strategy, where the orientation of the parallel scanning vectors is rotated by 90° after each layer. To investigate the effect of a modified scanning pattern, a wall geometry with a height H of 15.0 mm, a length L of 14.0 mm, and a width W of 7.0 mm, as illustrated in Figure 1 (right), was selected. For this wall geometry, the conventional 0°/90° and a modified 80°/100° bidirectional raster scanning strategy were applied. Table 2 details the experimental conditions for the specimen fabrication.

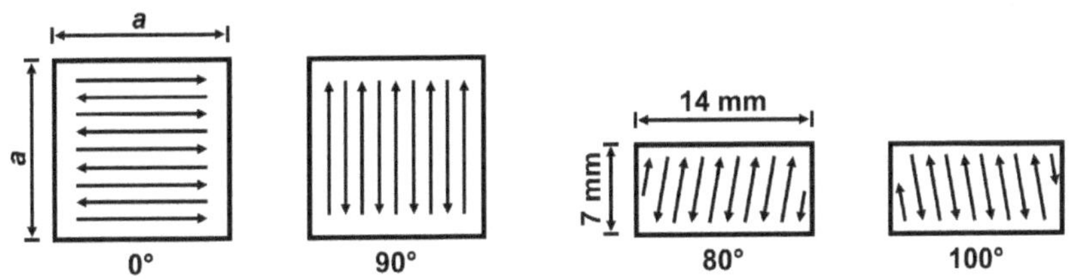

Figure 1. Specimen geometries and scanning vector orientations.

Table 2. Experimental conditions for the specimen fabrication.

Specimen	$L \times W \times H$ [mm^3]	Parameter Set [-]	Scanning [-]	Substr. Temp. [°C]
S1	5 × 5 × 5	P1	0°/90°	20
S2	5 × 5 × 5	P2	0°/90°	20
S3	5 × 5 × 5	P3	0°/90°	20
S4	5 × 5 × 5	P4	0°/90°	20
S5	10 × 10 × 10	P1	0°/90°	20
S6	10 × 10 × 10	P2	0°/90°	20
S7	10 × 10 × 10	P3	0°/90°	20
S8	10 × 10 × 10	P4	0°/90°	20
S9	15 × 15 × 15	P1	0°/90°	20
S10	15 × 15 × 15	P2	0°/90°	20
S11	15 × 15 × 15	P3	0°/90°	20
S12	15 × 15 × 15	P4	0°/90°	20
S13	15 × 15 × 15	P3	0°/90°	975
S14	14 × 7 × 15	P3	0°/90°	20
S15	14 × 7 × 15	P3	0°/90°	975
S16	14 × 7 × 15	P3	80°/100°	20
S17	14 × 7 × 15	P3	80°/100°	975

Table 3 lists the selected process parameter combinations P1 to P4. The laser power was set constant to 1000 W, and scanning speed and powder flow rate were varied on two levels each. These parameters were selected based on previous unpublished work of the authors, where comparable low scanning speeds and powder flow rates led to lowest porosity. For all specimens, the hatching distance d between the scanning vectors was 1.1 mm and the theoretical layer height Δz was between 0.54 and 0.84 mm, based on single layer experiments with the respective process parameter combinations.

Table 3. Processing parameter sets.

Parameter Set	Laser Power [W]	Scanning Speed [mm/min]	Powder Flow Rate [g/min]
P1	1000	250	2.0
P2	1000	300	2.0
P3	1000	250	2.5
P4	1000	300	2.5

For a variation of the initial substrate temperature, a laser preheating process was applied as illustrated in Figure 2. The working distance was raised to 50 mm and the scanning speed to 3000 mm/min to avoid local substrate melting. The machine table was tilted to 40° to reduce back reflection, and a rectangular scanning pattern was followed for 50 repetitions and a duration of 100 s. For the final 10 repetitions, the powder feeding was activated to obtain a stable material flow before the deposition started. An Optris PI 1M infrared camera (Optris GmbH, Berlin, Germany) with a spectral range between 0.85 µm to 1.1 µm was used to measure the substrate temperature during the developed preheating process.

Metallographic preparation and image analysis were conducted for the quantitative assessment of the cracking phenomena. The specimens were cut with a water-cooled abrasive grinding disc, embedded in thermoplastic resin, ground with SiC foils up to grit 4000, and polished with 3 µm and 1 µm diamond suspension. In order to fully reveal the cracks, an etchant consisting of 100 mL H_2O, 100 mL HCl, and 5 g Cu_2Cl (Kalling II) was applied. Full cross-section images were taken on a Keyence VHX 5000 digital microscope (Keyence, Osaka, Japan) in both polished and etched condition, and the length of each crack was analyzed using ImageJ software (version 1.51p, ImageJ, National Institutes of Health, Bethesda, MD, USA). The measurement uncertainty for the quantitative crack analysis was assumed to be 5%. A selected specimen was examined with a Zeiss EVO 10 scanning electron microscope (SEM, Carl Zeiss AG, Jena, Germany) equipped with an Oxford Instruments X-Max^N detector for energy-dispersive X-ray spectroscopy (EDS, Oxford Instruments, Abingdon, UK).

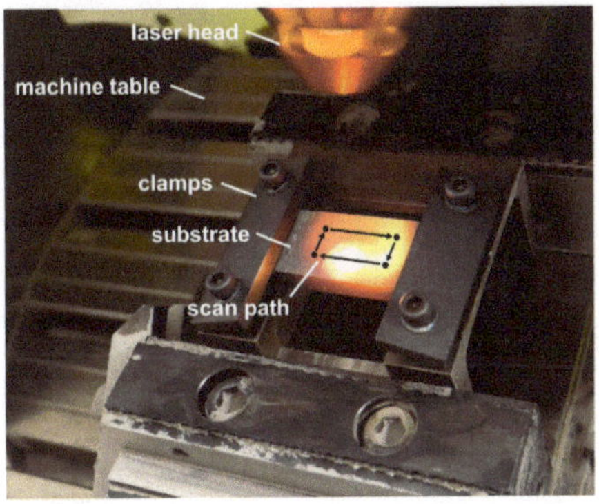

Figure 2. Experimental setup for the laser preheating process.

3. Results and Discussion

Figure 3 displays the crack densities for varying cube sizes and processing parameters. The crack density significantly increases for larger specimen geometries. While the specimens with the smallest edge length a = 5 mm are nearly crack-free, the crack densities for the larger cubes rise to between 0.08 and 0.20 mm/mm^2 for a = 10 mm and to between 0.45 and 0.57 mm/mm^2 for a = 15 mm. This corresponds to an average increase in crack density by 348% from edge length a = 10 mm to a = 15 mm. This finding may be of interest for repair applications, where the deposit size may be influenced by removing more or less material around the area of the defect prior to the deposition process.

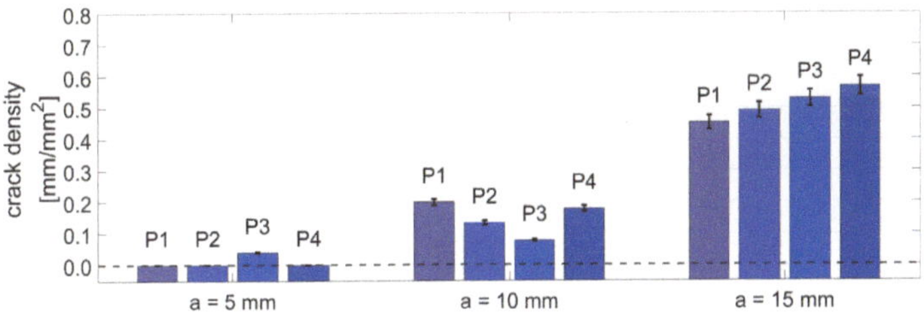

Figure 3. Crack density for specimens S1 to S12 with varied cube geometry and process parameter sets P1 to P4. The initial substrate temperature was constant around 20 °C. Error bars represent 5% measurement uncertainty.

There is no consistent tendency regarding the effect of processing parameters. The volumetric energy input

$$e_v = \frac{P}{vd\Delta z} \quad (1)$$

with laser power P, scanning speed v, hatching distance d, and theoretical layer height Δz for the tested conditions varied only between 270 and 374 J/mm^3, which presumably led to similar melt pool and thermal boundary conditions. Therefore, the narrow levels of the selected deposition parameters seem to inhibit a clear statement about the process parameter influence. In contrast, the specimen geometry significantly affects both the melt pool and thermal boundary conditions. For smaller specimens, a larger amount of heat accumulation occurs, because of the shorter scanning vectors and the lower amount of conductive heat transfer through the already deposited material. This heat accumulation may prevent cracking by three mechanisms: first, as discussed by Grange et al. [21], the healing of cracks by the deposition of the following layer is supported by melt pools that are larger and longer in the liquid state. Second, the increased base temperature leads to lower cooling rates during solidification and less segregation of low-melting phases at the grain boundaries, which is, according to Chen et al. [33], especially important for the prevention of liquation cracks. Third, as described by Stueber et al. [27], the lower cooling rates result in a lower magnitude of shrinking stresses, which is effective for the reduction of all types of hot cracking modes in nickel-based super alloys. According to these findings, there seems to be the potential to further reduce the crack densities of as-deposited alloy 738LC by increased substrate temperatures and optimized scanning strategies with shorter scanning vector lengths.

Figure 4 shows the temperature of the substrate material during the developed laser preheating process determined by three measurements with an infrared camera. The substrate temperature increases to above 1100 °C and exponentially decays after preheating for 102.75 s when the laser is turned off. For the following experiments, the deposition starts 5.70 s after the end of preheating. Therefore, the initial substrate temperature before the deposition start can be determined from the temperature profiles at a processing time of 108.45 s. The determined substrate temperatures at this time for the three tests are 967.1 °C, 976.7 °C, and 984.3 °C, respectively. Hence, an average temperature of 975 °C can be indicated as the initial substrate temperature for the following experiments with laser preheating.

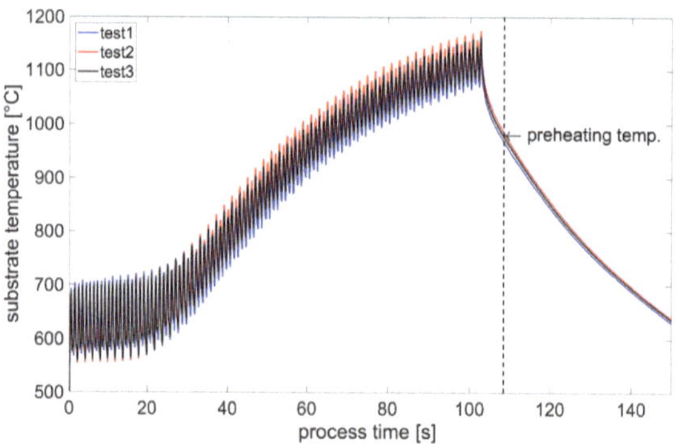

Figure 4. Temperature profiles of the developed laser preheating process. The initial substrate temperature for the following experiments was determined at a processing time of 108.45 s, as indicated by the dashed line.

Figure 5 illustrates the effect of initial substrate temperatures and scanning strategies on the resulting crack densities for two different specimen geometries. For the cuboid geometry with an edge length a = 15 mm, laser preheating of the substrate reduces the crack density to 0.10 mm/mm^2, compared to 0.53 mm/mm^2 for the deposition at room temperature. For the wall geometry with a cross section of 14.0 mm × 7.0 mm, the scanning vector length is 14.0 mm and 7.0 mm for orientations of 0° and 90°, respectively. For the modified 80°/100° scanning strategy, as visualized in Figure 1 (right), the maximum vector length is shortened to 7.1 mm, which is expected to reduce the amount of residual stress. However, the measured crack density without preheating for both scanning strategies has the same value of 0.08 mm/mm^2. Hence, the residual stress for both conditions is larger than the local material strength, and the optimized deposition pattern does not prevent cracks for this geometry. In contrast, the application of the laser preheating process to the wall geometry led to crack-free structures, presumably due to the lower levels of shrinkage stresses as described by Stueber et al. [27]. Therefore, laser preheating has a larger potential for crack prevention compared to process parameter and scanning strategy optimizations because it reduces residual stress to a greater extend.

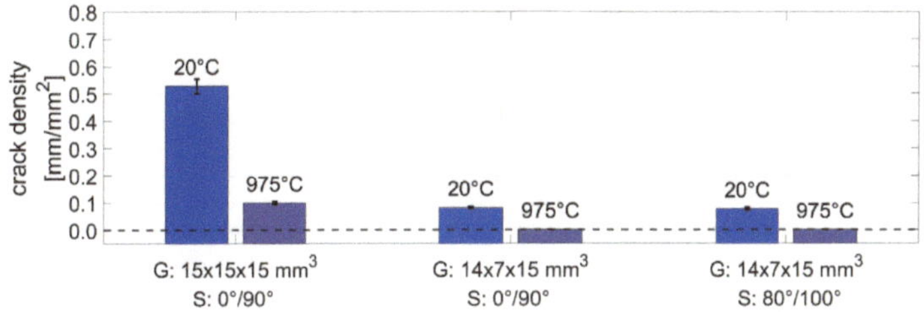

Figure 5. Crack density for different specimen geometries G, scanning strategies S, and initial substrate temperatures. Error bars represent 5% measurement uncertainty.

Figure 6 compares the cross section and crack density analysis for the cuboid geometries with different initial substrate temperatures (specimens S11 and S13). Without laser preheating, a large number of cracks appear throughout the specimen, except for some regions in the bottom and the top layer. In the bottom layer, the lower crack density may be related to the dilution with the substrate material 1.4404, which may lead to a smaller amount of Al- and Ti-enriched low-melting phases. In the top layer, the absence of large cracks strongly indicates that the main cracking mechanism is liquation cracking, which occurs in the heat-affected zone of the already deposited material underneath the latest deposited layer with a high potential for crack propagation over several layers. For the specimen with preheated substrate, there is a crack-free zone in the lower half of the structure. In this zone, the preheating process led to the highest initial substrate temperatures and lowest cooling rates. Hence, the amount of residual stress and the segregation of low-melting phases were reduced to the greatest extend in this lower region. In the upper regions, heat losses due to conduction, convection, and radiation during the process gradually reduced the temperature of the substrate and deposited structure over time. Consequently, residual stress and segregation increased, such that crack formation could not be completely prevented for this specimen geometry. This is similar to the findings of Chauvet et al. [20] for the deposition of a γ'-hardenable nickel alloy by electron beam powder bed fusion (PBF-EB) with preheating, where the 30 mm high test structure was crack-free only for the first 12 mm. Therefore, an initial preheating process alone may prevent hot cracking only for small specimen geometries. Further reduction of the crack density for larger structures may be possible by intermediate preheating to maintain a higher temperature of the already deposited structure. Alternatively, a laser remelting process as described by Leech [34] and as developed in a previous work by the authors [35] may be applied in between each layer.

Figure 7 illustrates the microstructure and the details of specimen S13 in etched condition. The microstructure shows a pattern of directional solidification and large, elongated grains. This is similar to the findings of Xu et al. [28] for preheating temperatures of 800 °C and 900 °C, where hot cracking could not be prevented, whereas for a preheating temperature of 1050 °C the fraction of equiaxed grains significantly increased and the deposit was free of cracks. The magnification of the crack in Figure 7 shows that the crack is located along the boundaries of multiple grains. Hence, grain boundary liquation may have caused the formation of this crack.

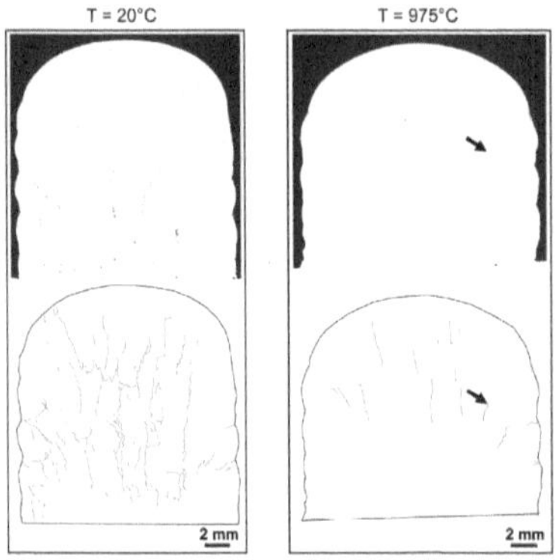

Figure 6. Polished cross-sections (**top**) and crack visualization (**bottom**) for cubes with 15 mm edge length fabricated without (**left**) and with (**right**) laser preheating to increase the initial substrate temperature T. The arrows show the location of an exemplary crack that is illustrated at higher magnification in the following figure.

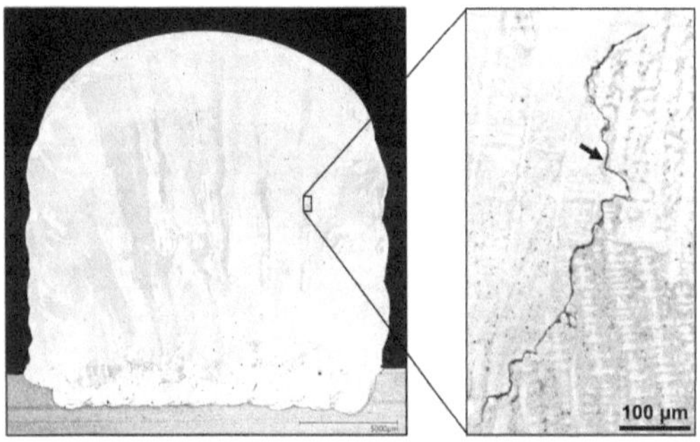

Figure 7. Optical microscope image of specimen S13. The arrow indicates the location that is shown at a higher magnification in Figure 8.

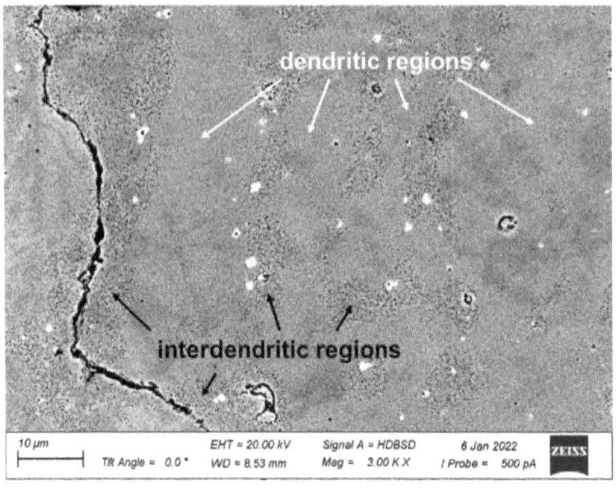

Figure 8. Backscatter electron image of a cracked grain boundary and its surroundings.

Figure 8 shows a backscatter electron image of a section of the previously shown crack and its surrounding dendritic microstructure. Within the interdendritic regions, block-shaped particles with a size of between 0.2 and 2.0 µm are present that appear white in the image. Energy-dispersive X-ray spectroscopy (EDS) revealed an enrichment of Ta and Ti in the interdendritic regions and identified the white particles as mixed carbides with enhanced contents of C, Cr, Ta, Ti, and W. As described by Ojo et al. [36], the high susceptibility to liquation cracking of alloy 738 is related to the liquation of γ' precipitates, and resolidified γ-γ' eutectic products may be found in the crack vicinity. Similarly, Chen et al. [33] confirmed the presence of γ-γ' colonies in the HAZ of alloy 738 welds. Hence, the enrichment of Ti in the interdentritic regions at the crack surroundings indicate that grain boundary liquation may be a possible explanation for crack formation in the present study.

Figure 9 displays further details of a cracked area at higher magnification, and Table 4 lists the chemical composition detected by EDS for selected points of interest. Due to the inaccuracy of EDS for low atomic weights, the measured contents of carbon and oxygen may significantly deviate from the actual content of the specimen and are for comparison purposes only. Point 1 represents the bulk material, where the chemical composition is similar to the raw powder material. The enrichment in C, Nb, Ta, Ti, and W identifies the white particles (Point 2 and 3) as mixed carbides. Within the crack, increased levels of oxygen are presumed to be associated with oxides deposited in the terminal stage of solidification or residual components from the metallographic preparation.

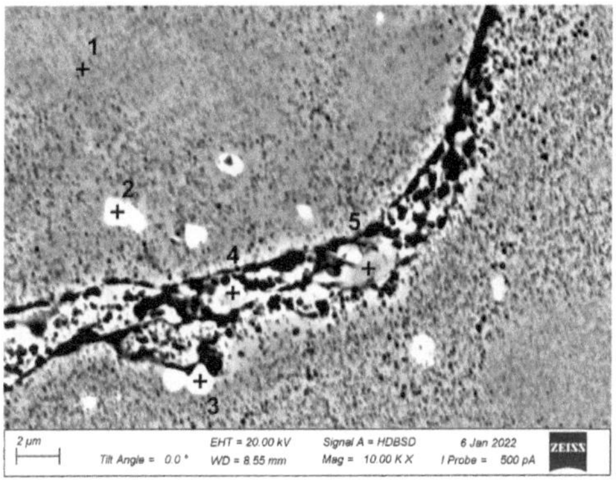

Figure 9. Backscatter electron image of a cracked region and selected points for EDS analysis.

Table 4. EDS results in wt. % of the points illustrated in Figure 8.

Element	Al	C	Co	Cr	Nb	Ni	O	Ta	Ti	W
Point 1	3.1	4.4	8.4	16	0.5	57.6	0	1.9	2.6	3.3
Point 2	0.3	12.5	1.8	4.6	9.3	11.0	0	25.5	25.3	6.0
Point 3	2.6	9.0	2.6	6.1	10.3	18.3	7.1	21.1	14.3	4.6
Point 4	3.1	5.0	7.8	15.2	0.8	55.5	2.2	2.1	3.3	3.0
Point 5	3.0	4.6	7.7	15.1	0.7	55.5	2.9	2.3	3.5	2.7

The main limitations of this study are related to the statistical uncertainty of the experimental results as each condition was measured only once and only one type of material was used. Since each process parameter combination was fabricated and analyzed only once, no statement about the repeatability of the measurements is possible. Hence, future studies may include the analysis of multiple specimens for each condition for the statistical validation of the results. Furthermore, future studies may validate laser preheating and DMD for other brittle materials, such as Mo-Si-B based alloys, for which Higashi and Yoshimi [37] showed a high susceptibility to crack formation.

4. Conclusions

Within this study, a laser preheating process was developed for direct metal deposition (DMD) of precipitation-hardenable nickel-based super alloys. The process was successfully validated for Inconel 738LC, and the main conclusions are as follows:

- The crack formation significantly depends on the size of the deposited structure and the solidification conditions because smaller structures heat up and experience low levels of thermal stress. While the smallest cubes with 5 mm edge length were nearly free of cracks, the crack density increased from between 0.08 and 0.20 mm/mm^2 to between 0.45 and 0.57 mm/mm^2 for the edge lengths 10 mm and 15 mm, respectively.
- The variation of scanning speed, powder flow rate, and scanning pattern did not significantly affect the crack formation within the analyzed conditions of this study.
- Laser preheating strongly reduced the crack density for all tested conditions. Structures of dimension 14 mm × 7 mm × 15 mm built on a substrate at room temperature showed a crack density of 0.08 mm/mm^2, while the deposition on preheated substrates led to microstructures without detected cracks. For the largest analyzed specimens, laser preheating led to a reduction in crack density by a factor of 5.3.
- The proposed process is implementable on existing DMD machines and does not require any additional hardware. Therefore, the laser preheating process appears to be economically feasible for the repair of super alloy components by DMD.

Author Contributions: Conceptualization, F.S. and K.W.; methodology, F.S. and K.P.; validation, F.S. and K.P.; formal analysis, F.S.; investigation, F.S. and K.P.; resources, M.B. and K.W.; writing—original draft preparation, F.S.; writing—review and editing, K.P., M.B. and K.W.; visualization, F.S.; supervision, M.B. and K.W.; project administration, F.S. and K.W.; funding acquisition, M.B. and K.W. All authors have read and agreed to the published version of the manuscript.

Funding: The authors would like to acknowledge the support by the funding agency Innosuisse (grant number 25498) and the companies GF Machining Solutions, GF Precicast, and ABB Schweiz AG, Turbocharging.

Institutional Review Board Statement: Not applicable.

Informed Consent Statement: Not applicable.

Data Availability Statement: The datasets generated during and/or analyzed during the current study are available from the corresponding author on reasonable request.

Acknowledgments: The authors would like to thank Knut Krieger, Daniel Eisenbarth, and Arthur Riesen for their assistance within this study.

Conflicts of Interest: The authors declare no conflict of interest.

References

1. DebRoy, T.; Wei, H.L.; Zuback, J.S.; Mukherjee, T.; Elmer, J.W.; Milewski, J.O.; Beese, A.M.; Wilson-Heid, A.; De, A.; Zhang, W. Additive manufacturing of metallic components—Process, structure and properties. *Prog. Mater. Sci.* **2018**, *92*, 112–224. [CrossRef]
2. Xiong, Z.; Li, Z.; Sun, Z.; Hao, S.; Yang, Y.; Li, M.; Song, C.; Qiu, P.; Cui, L. Selective laser melting of NiTi alloy with superior tensile property and shape memory effect. *J. Mater. Sci. Tech.* **2019**, *35*, 2238–2242. [CrossRef]
3. Elahinia, M.; Shayesteh Moghaddam, N.; Taheri Andani, M.; Amerinatanzi, A.; Bimber, B.A.; Hamilton, R.F. Fabrication of NiTi through additive manufacturing: A review. *Prog. Mater. Sci.* **2016**, *83*, 630–663. [CrossRef]
4. Smelov, V.G.; Sotov, A.V.; Murzin, S.P. Particularly Selective Sintering of Metal Powders by Pulsed Laser Radiation. *Key Eng. Mater.* **2016**, *685*, 403–407. [CrossRef]
5. Mazumder, J. Laser-aided direct metal deposition of metals and alloys. In *Laser Additive Manufacturing*; Woodhead Publishing: Sawston, UK, 2017; pp. 21–53. [CrossRef]
6. Wirth, F.; Wegener, K. A physical modeling and predictive simulation of the laser cladding process. *Addit. Manuf.* **2018**, *22*, 307–319. [CrossRef]
7. Wirth, F.; Arpagaus, S.; Wegener, K. Analysis of melt pool dynamics in laser cladding and direct metal deposition by automated high-speed camera image evaluation. *Addit. Manuf.* **2018**, *21*, 369–382. [CrossRef]
8. Rettberg, R.; Kraenzler, T. Hybrid manufacturing: A new additive manufacturing approach for closed pump impellers. In *Industrializing Additive Manufacturing*; Springer: Cham, Switzerland, 2021; pp. 146–159. [CrossRef]
9. Hosseini, E.; Popovich, V.A. A review of mechanical properties of additively manufactured Inconel 718. *Addit. Manuf.* **2019**, *30*, 100877. [CrossRef]
10. Murzin, S.P.; Kazanskiy, N.L.; Stiglbrunner, C. Analysis of the Advantages of Laser Processing of Aerospace Materials Using Diffractive Optics. *Metals* **2021**, *11*, 963. [CrossRef]

11. Zhong, M.; Sun, H.; Liu, W.; Zhu, X.; He, J. Boundary liquation and interface cracking characterization in laser deposition of Inconel 738 on directionally solidified Ni-based superalloy. *Scr. Mater.* **2005**, *53*, 159–164. [CrossRef]
12. Montazeri, M.; Ghaini, F.M. The liquation cracking behavior of IN738LC superalloy during low power Nd:YAG pulsed laser welding. *Mater. Charact.* **2012**, *67*, 65–73. [CrossRef]
13. Rickenbacher, L.; Etter, T.; Hövel, S.; Wegener, K. High temperature material properties of IN738LC processed by selective laser melting (SLM) technology. *Rapid Prototyp. J.* **2013**, *19*, 282–290. [CrossRef]
14. Adegoke, O.; Andersson, J.; Brodin, H.; Pederson, R. Influence of Laser Powder Bed Fusion Process Parameters on Voids, Cracks, and Microhardness of Nickel-Based Superalloy Alloy 247LC. *Materials* **2020**, *13*, 3770. [CrossRef] [PubMed]
15. Liu, S.; Yu, H.; Wang, Y.; Zhang, X.; Li, J.; Chen, S.; Liu, C. Cracking, Microstructure and Tribological Properties of Laser Formed and Remelted K417G Ni-Based Superalloy. *Coatings* **2019**, *9*, 71. [CrossRef]
16. Liu, D.; Lippold, J.C.; Li, J.; Rohklin, S.R.; Vollbrecht, J.; Grylls, R. Laser Engineered Net Shape (LENS) Technology for the Repair of Ni-Base Superalloy Turbine Components. *Metall. Mater. Trans. A* **2014**, *45*, 4454–4469. [CrossRef]
17. DebRoy, T.; Mukherjee, T.; Milewski, J.O.; Elmer, J.W.; Ribic, B.; Blecher, J.J.; Zhang, W. Scientific, technological and economic issues in metal printing and their solutions. *Nat. Mater.* **2019**, *18*, 1026–1032. [CrossRef] [PubMed]
18. Kou, S. Predicting Susceptibility to Solidification Cracking and Liquation Cracking by CALPHAD. *Metals* **2021**, *11*, 1442. [CrossRef]
19. Xu, J.; Lin, X.; Guo, P.; Dong, H.; Wen, X.; Li, Q.; Xue, L.; Huang, W. The initiation and propagation mechanism of the overlapping zone cracking during laser solid forming of IN-738LC superalloy. *J. Alloys Compd.* **2018**, *749*, 859–870. [CrossRef]
20. Chauvet, E.; Kontis, P.; Jägle, E.A.; Gault, B.; Raabe, D.; Tassin, C.; Blandin, J.-J.; Dendievel, R.; Vayre, B.; Abed, S.; et al. Hot cracking mechanism affecting a non-weldable Ni-based superalloy produced by selective electron Beam Melting. *Acta. Mater.* **2018**, *142*, 82–94. [CrossRef]
21. Grange, D.; Bartout, J.D.; Macquaire, B.; Colin, C. Processing a non-weldable nickel-base superalloy by Selective Laser Melting: Role of the shape and size of the melt pools on solidification cracking. *Materialia* **2020**, *12*, 100686. [CrossRef]
22. Seidel, A.; Finaske, T.; Straubel, A.; Wendrock, H.; Maiwald, T.; Riede, M.; Lopez, E.; Brueckner, F.; Leyens, C. Additive Manufacturing of Powdery Ni-Based Superalloys Mar-M-247 and CM 247 LC in Hybrid Laser Metal Deposition. *Metall. Mater. Trans. A* **2018**, *49*, 3812–3830. [CrossRef]
23. Engeli, R.; Etter, T.; Hövel, S.; Wegener, K. Processability of different IN738LC powder batches by selective laser melting. *J. Mater. Process. Tech.* **2016**, *229*, 484–491. [CrossRef]
24. Engeli, R.; Etter, T.; Meidani, H. Gamma Prime Precipitation Strengthened Nickel-Base Superalloy for Use in Powder Based Additive Manufacturing process. U.S. Patent US 2017/0021453 A1, 26 January 2017.
25. Griffiths, S.; Ghasemi Tabasi, H.; Ivas, T.; Maeder, X.; De Luca, A.; Zweiacker, K.; Wróbel, R.; Jhabvala, J.; Logé, R.E.; Leinenbach, C. Combining alloy and process modification for micro-crack mitigation in an additively manufactured Ni-base superalloy. *Addit. Manuf.* **2020**, *36*, 101443. [CrossRef]
26. Chiang, M.F.; Chen, C. Induction-assisted laser welding of IN-738 nickel–base superalloy. *Mater. Chem. Phys.* **2009**, *114*, 415–419. [CrossRef]
27. Stueber, R.; Milidantri, T.; Tadayon, M. Welding High-Strength Nickel Base Superalloys. U.S. Patent US5374319A, 28 September 1990.
28. Xu, J.; Lin, X.; Guo, P.; Hu, Y.; Wen, X.; Xue, L.; Liu, J.; Huang, W. The effect of preheating on microstructure and mechanical properties of laser solid forming IN-738LC alloy. *Mat. Sci. Eng. A* **2017**, *691*, 71–80. [CrossRef]
29. Cloots, M.; Uggowitzer, P.J.; Wegener, K. Investigations on the microstructure and crack formation of IN738LC samples processed by selective laser melting using Gaussian and doughnut profiles. *Mater. Des.* **2016**, *89*, 770–784. [CrossRef]
30. Ramakrishnan, A.; Dinda, G.P. Direct laser metal deposition of Inconel 738. *Mat. Sci. Eng. A* **2019**, *740–741*, 1–13. [CrossRef]
31. Lee, Y.S.; Kirka, M.M.; Ferguson, J.; Paquit, V.C. Correlations of cracking with scan strategy and build geometry in electron beam powder bed additive manufacturing. *Addit. Manuf.* **2020**, *32*, 101031. [CrossRef]
32. Cortina, M.; Arrizubieta, J.I.; Ruiz, J.E.; Ukar, E.; Lamikiz, A. Latest Developments in Industrial Hybrid Machine Tools that Combine Additive and Subtractive Operations. *Materials* **2018**, *11*, 2583. [CrossRef]
33. Chen, K.-C.; Chen, T.-C.; Shiue, R.-K.; Tsay, L.-W. Liquation Cracking in the Heat-Affected Zone of IN738 Superalloy Weld. *Metals* **2018**, *8*, 387. [CrossRef]
34. Leech, P.W. Laser surface melting of a complex high alloy steel. *Mat. Des.* **2014**, *54*, 539–543. [CrossRef]
35. Soffel, F.; Lin, Y.; Keller, D.; Egorov, S.; Wegener, K. Laser Remelting Process Simulation and Optimization for Additive Manufacturing of Nickel-Based Super Alloys. *Materials* **2022**, *15*, 177. [CrossRef] [PubMed]
36. Ojo, O.A.; Richards, N.L.; Chaturvedi, M.C. Contribution of constitutional liquation of gamma prime precipitate to weld HAZ cracking of cast Inconel 738 superalloy. *Scr. Mater.* **2004**, *50*, 641–646. [CrossRef]
37. Higashi, M.; Yoshimi, K. Electron beam surface melting of MoSiBTiC alloys: Effect of preheating on cracking behavior and microstructure evolution. *Mater. Des.* **2021**, *209*, 110010. [CrossRef]

metals MDPI

Article

Hydrogen Embrittlement of Inconel 718 Manufactured by Laser Powder Bed Fusion Using Sustainable Feedstock: Effect of Heat Treatment and Microstructural Anisotropy

Naveen Karuthodi Mohandas [1,*], Alex Giorgini [2], Matteo Vanazzi [2], Ton Riemslag [1], Sean Paul Scott [1] and Vera Popovich [1,*]

1 Department of Materials Science and Engineering, Delft University of Technology, Mekelweg 2, 2628 CD Delft, The Netherlands
2 F3nice, Via Roccoli, 252, 23010 Piantedo, SO, Italy
* Correspondence: n.k.mohandas@tudelft.nl (N.K.M.); v.popovich@tudelft.nl (V.P.)

Abstract: This study investigated the in-situ gaseous (under 150 bar) hydrogen embrittlement behaviour of additively manufactured (AM) Inconel 718 produced from sustainable feedstock. Here, sustainable feedstock refers to the Inconel 718 powder produced by vacuum induction melting inert gas atomisation of failed printed parts or waste from CNC machining. All Inconel 718 samples, namely AM-as-processed, AM-heat-treated and conventional samples showed severe hydrogen embrittlement. Additionally, it was found that despite its higher yield strength (1462 ± 8 MPa) and the presence of δ phase, heat-treated AM Inconel 718 demonstrates 64% lower degree of hydrogen embrittlement compared to the wrought counterpart (Y.S. 1069 ± 4 MPa). This was linked to the anisotropic microstructure induced by the AM process, which was found to cause directional embrittlement unlike the wrought samples showing isotropic embrittlement. In conclusion, this study shows that AM Inconel 718 produced from recycled feedstock shows better hydrogen embrittlement resistance compared to the wrought sample. Furthermore, the unique anisotropic properties, seen in this study for Inconel 718 manufactured by laser powder bed fusion, could be considered further in component design to help minimise the degree of hydrogen embrittlement.

Keywords: Inconel 718; hydrogen embrittlement; additive manufacturing; recycled powder; laser powder bed fusion; anisotropy

1. Introduction

Laser powder bed fusion (L-PBF) is an additive manufacturing (AM) technology that has been gaining a lot of interest as it allows production of parts with complex geometries and eliminates expensive tooling [1]. Unlike the conventional manufacturing processes, where material is removed to make the final component, AM parts are manufactured layer-wise as per the required geometry. Thus, AM provides immense design flexibility and minimal material wastage [2].

In addition, L-PBF can further promote sustainability by recycling the feedstock. This can be done in two ways: one is to reuse the powder that has undergone multiple printing jobs [3] and the other is to use end-of-life products as raw material for powder production [4,5]. To make a clear distinction between the powder reused from printing and powder produced by recycling end-of-life products, the former is referred to as reused powder and the latter is referred to as recycled powder. The issue with reused powder is that it undergoes multiple thermal cycles during the L-PBF process, altering its properties. Numerous studies have underlined the influence of reusing powder feedstock on mechanical properties [3,6–10]. Most studies are, however, restricted to static properties. Sutton et al. [11] observed that though there was no variation in tensile properties, there was a steady decline in the impact toughness of austenitic stainless steel AISI 304L. Hence, further studies are needed to understand the influence of reused powder on dynamic and environmental properties. When it comes to recycled powder, only preliminary studies were found [4,5]. These preliminary studies showed promising trends for static properties; however, more extensive studies are needed for validation.

Inconel 718 is a nickel-based superalloy that is often used in turbine blades and heat exchangers where high performance in extreme environments is required [12]. The good welding properties of the alloy make it suitable for L-PBF technology. Soller et al. [13] describes how L-PBF could be utilised to produce injectors from Inconel 718 and stainless steel for a liquid rocket engine, which are complex and demanding. The AM technology has the potential to eliminate numerous integration steps involved in conventional manufacturing, thereby reducing production costs [2]. Nevertheless, the application of Inconel 718 for hydrogen fuel is challenging, as it is susceptible to hydrogen embrittlement (HE) [14–16]. Hydrogen embrittlement is defined as loss of ductility and strength in the presence of hydrogen [17,18]. The degree of embrittlement is dependent on the microstructure and environment. L-PBF has a potential application here as the microstructure can be tailored by controlling process parameters such as heat-source parameters and feedstock properties [19,20]. Apart from these, post-processing treatments such as heat treatments and hot isostatic pressing (HIP) also influence the microstructure and functional properties. Numerous studies have been performed to optimise these process and post-process parameters and achieve properties comparable or superior to its conventional counterparts [21–24].

The dominant mechanisms prevalent during the hydrogen embrittlement is a debated topic. The major controversy exists in whether hydrogen promotes brittle cleavage failure or ductile mode of failure [25,26]. Based on fracture surface observations, various mechanisms such as hydrogen-enhanced localised plasticity (HELP), hydrogen-enhanced decohesion (HEDE), adsorption-induced dislocation emission (AIDE) and others have been proposed [17,27]. In the case of Inconel 718, the presence of different phases (γ', γ'', δ and Laves phase) further complicates it. Additionally, Hicks et al. [28] reported that the mode of failure also depends on the concentration of hydrogen. Hence, the differences arising from the experimental methods and material compositions makes it difficult to understand and separate the underlying hydrogen embrittlement mechanisms. For Inconel 718, HELP and HEDE are the most commonly observed mechanisms [14,29].

Only a limited amount of literature is available for hydrogen embrittlement of L-PBF Inconel 718 [30–33]. The HE behaviour is found to be similar to that of the conventional counterpart, nevertheless, the degree of HE varies depending on the post-processing treatments. Hesketh et al. [33] studied the effect of part orientation on the influence of hydrogen in L-PBF produced samples after non-standard heat treatment (1030 °C for 1 h and aged at 783 °C for 6 h 45 m) and also compared with wrought specimens with identical heat treatment. The experiment was conducted with in-situ hydrogen charging by electrochemical means, and it was reported that the vertical L-PBF samples (tensile axis parallel to build direction) showed higher HE resistance compared to the horizontal sample (tensile axis perpendicular to build direction) which has a higher yield and tensile strength. This is due to the microstructural differences in the horizontal and vertical samples and the higher strength of horizontal samples leading to higher stress state. From the fractography, Hesketh concludes that the effect of porosity, found to be around 0.18%, is negligible. However, the effect of residual stress, which can lead to varying concentrations of hydrogen in the specimen, was not investigated. Hence, in order to understand hydrogen embrittlement in components produced from L-PBF, it is necessary to account for the process-specific properties and defects; and how it interacts with hydrogen. Additionally, to the best of authors knowledge there is no work regarding the effect of feedstock powder on HE of L-PBF Inconel 718.

In this study, the effect of recycled feedstock powder and post-process heat-treatments on hydrogen embrittlement of L-PBF Inconel 718 was investigated. Slow strain rate tensile tests in in-situ gaseous hydrogen environment and extensive microstructural characterization were performed on L-PBF Inconel 718 and compared with conventional counterparts.

2. Materials and Methods

In this research, Inconel 718 powder provided by F3nice [34] was used for the L-PBF process. The powder was produced from recycled materials, namely leftovers from CNC machining, failed printed parts and disqualified powder, using vacuum induction-melting inert gas atomization (VIGA) technology. The samples are printed with an AMCM M290 (customised EOS M290) machine equipped with 1070 nm wavelength Ytterbium fibre laser. A stripe scanning strategy with laser power of 285 W was used in combination with 960 mm/s scan speed and 0.11 mm hatch distance, leading to a volumetric energy density (VED) of 67 J/mm^3. To prevent oxidation during the L-PBF process, the samples were printed in an inert argon gas environment. The process parameters are summarised in Table 1. The recycled powder feedstock is in accordance to ASTM F3055 [35] and has a size distribution of 55.5 μm, 36 μm, and 23.2 μm corresponding to D_{90}, D_{50} and D_{10} particle size distribution, respectively. The SEM micrographs showing the particle morphology is given in Figure 1.

Table 1. Process parameters used for L-PBF of Inconel 718.

Laser Power (W)	Scan Speed (mm/s)	Hatch Distance (mm)	Layer Thickness (μm)	Volumetric Energy Density (VED) (J/mm^3)
285	960	80	40	67

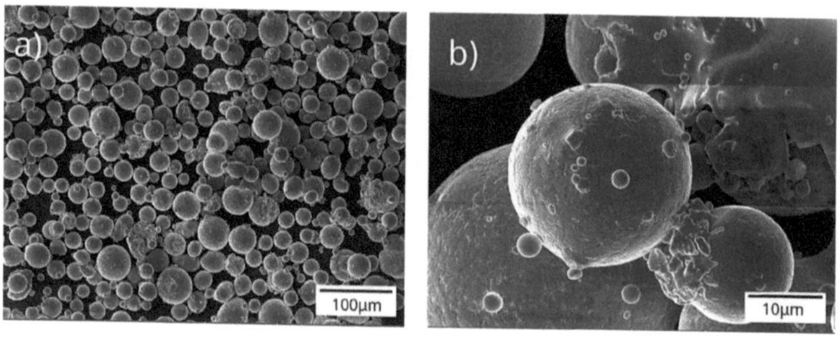

Figure 1. SEM micrographs of (**a**) recycled Inconel 718 powder particles, (**b**) high magnification image showing satellite particles on the surface.

To evaluate the worst-case scenario, the samples were printed in the horizontal direction (as shown in Figure 2). This is because the degree of hydrogen embrittlement (HE) is found to be higher in materials with higher yield strength [36]. In L-PBF, horizontally built samples are found to be more susceptible to (HE) than vertically built samples [33]. The AM samples studied herein are investigated in both as-processed and heat-treated conditions. In the remainder of the article, these samples are referred to as AM-AP and AM-HT, respectively. The heat treatment applied in this study involves solutionising at 980 °C for 1 h, followed by double ageing at 720 °C for 6 h in a furnace, then cooled to 620 °C and held for 10 h, which is summarised in Table 2.

Table 2. Post-processing heat treatment parameters used for the Inconel 718 samples.

Sample	Heat Treatment	Temperature	Holding Time	Cooling
Conventional	Solutionising Ageing	1032 °C 790 °C	1 h 6 h	WC AC
AM-HT *	Solutionising Ageing	980 °C 720 °C 620 °C	1 h 8 h 8 h	AC FC to 620 °C @ 55 °C/h AC

* As per AMS5663 [37].

The L-PBF Inconel 718 was compared with conventional hot-rolled samples. The conventional hot-rolled samples after heat treatment (Table 2) were supplied by VDM metals GmbH [38]. It represents commercially used Inconel 718 in the industries as per the ASTM B 637 standard [39].

For microstructural characterisation, the samples were polished to mirror finish and etched using Kallings reagent (5 g CuCl2, 100 mL HCl and 100 mL ethanol). The microstructure was then analysed using an Olympus optical microscope. To determine the morphology of the precipitates, a Jeol JSM IT-100 scanning electron microscope (SEM) equipped with energy-dispersive spectroscopy (EDS) was used.

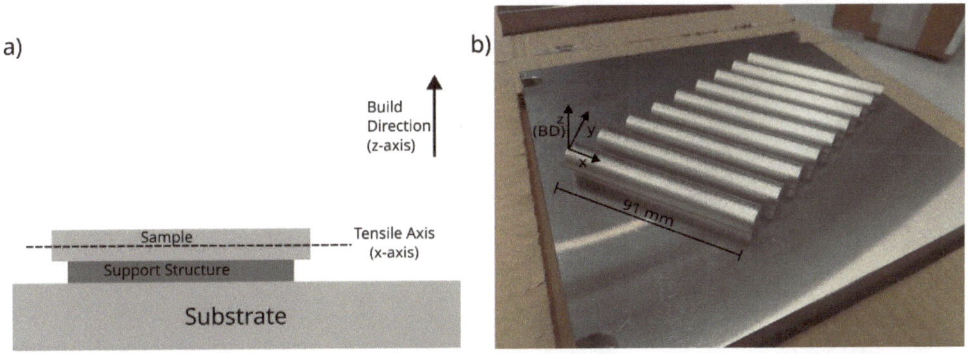

Figure 2. (**a**) Sample orientation (horizontal) for the L-PBF process, (**b**) AM samples. z-axis is the building direction (BD) and x-axis is the tensile axis.

For the AM samples, porosity measurements were additionally taken on the polished surface using a Keyence VHX-5000 digital microscope. A rectangular region of 2×2 mm^2 was defined and the porosity within this region was determined. The resolution of the pore size is limited to 1 µm. This was repeated in three different regions and the average porosity is calculated.

Vicker's hardness tests were performed on the samples as a preliminary estimate of mechanical properties. The hardness tests were performed on a Struers Durascan hardness tester with HV5 (5 kgf) scale to obtain the average hardness of the material. Five measurements were taken close to the centre of the sample in the direction perpendicular to the tensile axis (XY-plane in Figure 2), which is the build direction in the case of the AM samples.

X-ray diffraction analysis (XRD) was carried out to qualitatively determine the different phases precipitated in the conventional, AM and AM-HT samples. The XRD measurements were performed in a Bruker D8 Advance diffractometer with Co kα radiation. A step size of 0.035° 2θ with 45 kV and 40 mA was used. The peak identification was carried out in Bruker software Diffrac.EVA.

To characterise the effect of gaseous hydrogen on Inconel 718, the sample design developed by Boot et al. [40] (shown in Figure 3) for slow strain rate tensile (SSRT) test was used. This design has the advantage that the sample acts as the gas containment volume throughout the experiment, thus avoiding the need for autoclaves and sophisticated equipment. The AM-AP, AM-HT and conventional cylindrical samples were machined to the required dimensions for mechanical testing. A standard hydraulic 4-column load frame (MTS 311.21 Servo Hydraulic Load Frame) with a load capacity of 350 kN and modified adapters, as shown in Figure 3, was used to perform the in-situ experiments. The supply of gas during testing was provided through the top adapters connected to gas cylinders of hydrogen and nitrogen, each with a purity of 99%. The SSRT tests were performed with a cross head speed of 1.5 mm/h, which corresponded to a strain rate of ~10^{-5} s^{-1}. An Instron extensometer (model no: 2620-603) with 10 mm gauge length was then placed on the specimen over the notch. When a specimen began to yield during testing, the extensometer was removed to avoid damage during failure and then the crosshead displacement was used to determine the final strain in the material.

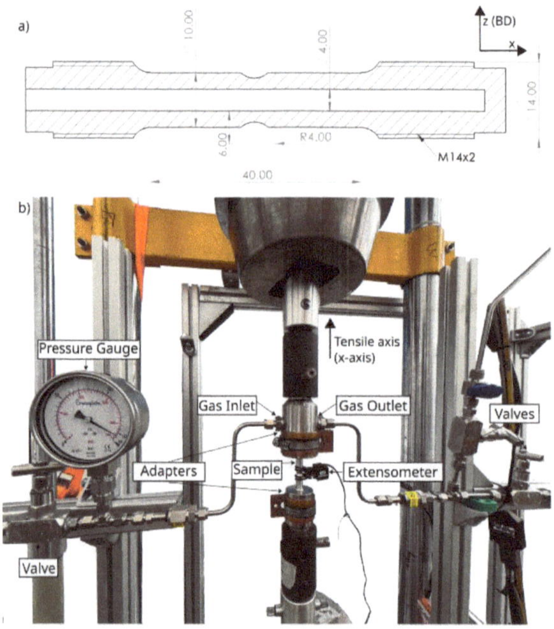

Figure 3. (**a**) Sample dimensions. (**b**) Test setup for SSRT. The tensile loading direction is the x-axis.

The SSRT tests were performed in both nitrogen (N_2) and hydrogen (H_2) environments, with a pressure of 150 bar at 25 °C. Two repetitions were performed for each testing condition for repeatability. For the hydrogen tests, the samples were additionally charged for 48 h before the test to allow for hydrogen diffusion within the sample.

After the SSRT tests, the area of cross section after failure (A_f) is determined at the fracture location using Keyence VHX-5000 digital microscope. The percentage reduction in area (RA) is found by comparing it with the initial cross sectional area (A_0) [40]:

$$\%RA = (A_0 - A_f)/A_0 \times 100\% \quad (1)$$

The degree of hydrogen embrittlement is measured as the relative reduction of area of samples tested in nitrogen ($\%RA_{N2}$) and hydrogen (RA_{H2}) environment [40]:

$$HEI = (\%RA_{N2} - \%RA_{H2})/\%RA_{N2} \times 100\% \quad (2)$$

where HEI is the hydrogen embrittlement index, representing the percentage of ductility that was lost due to the presence of hydrogen.

3. Results and Discussion

3.1. Microstructural Characterisation

The composition of the conventional and AM samples, as measured by X-ray fluorescence spectroscopy, is given in Table 3. A slight variation in the composition of the major alloying elements is noted, but is within the allowed limits as per ASTM B637 standard [39]. The elements Co and Cu are not detected in the AM samples; these elements are added to impart solid solution strengthening in Ni-based superalloys, and are not expected to majorly influence the precipitation of phases in this study [41].

Table 3. Composition (wt. %) of Inconel 718 obtained from X-ray fluorescence spectroscopy (XRF).

Sample	Ni	Cr	Fe	Nb	Mo	Ti	Al	Si	Mn	Mg	V	Co	Cu
Conventional	54.90	17.55	17.55	4.66	2.86	0.94	0.45	0.09	0.09	-	0.03	0.37	0.15
AM	54.84	18.25	17.40	4.71	3.01	1.09	0.36	0.12	0.11	0.05	0.02	-	-

Figure 4 shows the microstructure of the conventional Inconel 718 sample after heat treatment with equiaxed grains and annealing twins. During the heat treatment, the solutionising temperature of 1032 °C is higher than the δ solvus temperature (~1000 °C) [12], resulting in the dissolution of the δ phase. Cai et al. [42] reported that after 1 h of holding time at 1020 °C the δ phase fully dissolves in the matrix. In the current study at high magnification (Figure 4b), precipitates of δ phase with a size of ~0.8 µm are seen along the grain boundaries. Nevertheless, the fraction of the δ phase observed is small, as it is found only in certain regions and not throughout the grain boundaries. The γ' and γ'' strengthening phases form during the subsequent ageing heat treatment at 790 °C for 6 h. As these strengthening phases precipitate in the nanometre range, they are hard to discern in a scanning electron microscope.

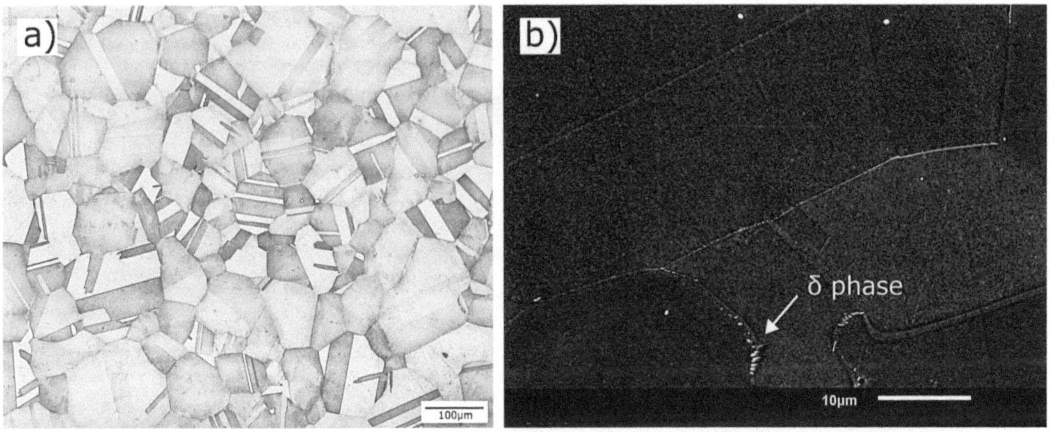

Figure 4. (**a**) Optical image of conventional Inconel 718 showing equiaxed grains. (**b**) SEM image showing δ phase.

Along the building direction, the AM-AP sample shows fish-scale-like morphology (Figure 5a), arising from the remelting of previous layers during layer-wise deposition of material. Due to the presence of solidified layers, grain growth in L-PBF occurs epitaxially; this is seen as dendritic structures spanning multiple melt pools in the microstructure [20]. Figure 5b) shows the cellular structures of the columnar dendrites. Laves phase (marked by arrows) forms in the interdendritic regions due to the micro-segregation of Nb and Ti during solidification [43]. As these segregations are detrimental to mechanical properties, post-processing heat treatments are recommended [20].

For AM-HT samples, solutionising the temperature (980 °C) and time (1 h) during the SA heat treatment is insufficient for complete recrystallisation. This is evident from the microstructure shown in Figure 6, where partial melt pool boundaries are still visible. In addition, at 980 °C only partial dissolution of Laves phase occurs and diffusion of Nb is inadequate for homogenisation. The partial dissolution of Laves phase leads to an increase in the concentration of Nb in the neighbourhood. This results in the formation of the acicular delta phase with a size of ~1.1 µm along the grain boundaries, as seen in (Figure 6b). This is in agreement with the observations of Chlebus et al. [44], Li et al. [43] and Zhang et al. [45]. A two-step ageing heat treatment is also performed for the precipitation and coarsening of γ' and γ'' strengthening phases. It is found that SA heat treatment does not eliminate the anisotropy in microstructure entirely, as complete recrystallisation does not occur. Sabelkin et al. [46] recommends a non-traditional heat treatment, which involves annealing at 1160 °C for 4 h to allow complete recrystallisation and anisotropy elimination.

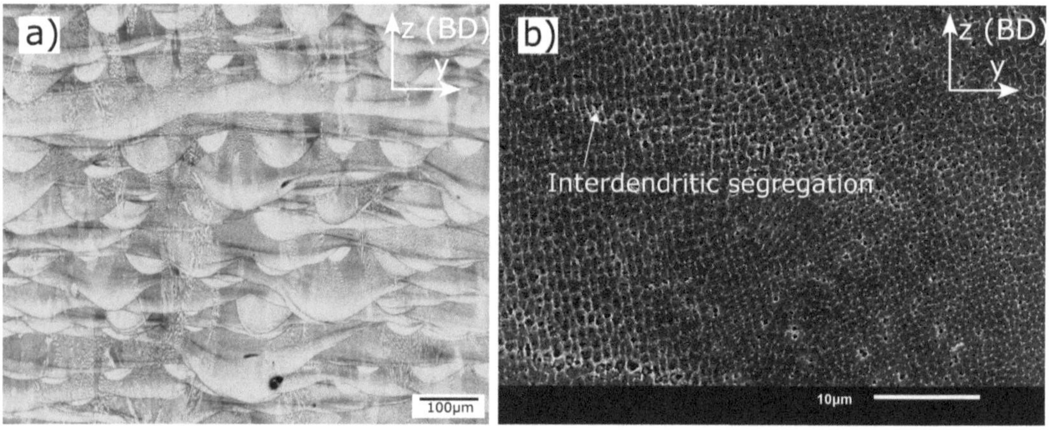

Figure 5. (**a**) Meltpool boundaries of AM-AP samples. (**b**) Cellular substructures with segregations along the boundaries in yz-plane.

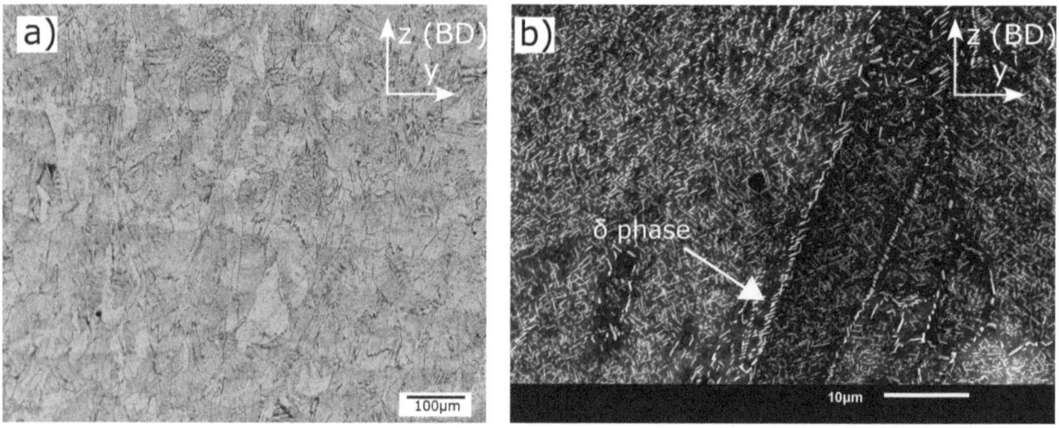

Figure 6. (**a**) Optical microscopy and (**b**) SEM in yz-plane showing microstructure of AM-HT samples.

The microstructures of AM-AP and AM-HT, perpendicular to the build direction (xy-plane), are shown in Figure 7. Scan tracks representative of the scanning strategy are visible in the AM-AP samples. Wang et al. [47] performed EBSD analysis and reported that the grains appeared equiaxed in this plane. Popovich et al. [19,48] performed similar studies for functionally graded Inconel 718 and reported that laser power strongly influenced texture, in turn, affecting anisotropy in samples. It was also found that HIP treatment promoted recrystallisation, whereas annealing heat treatment at 850 °C did not. This supports the observation in Figure 7b) vaguely showing melt pools (marked by dotted line) after only heat treatment.

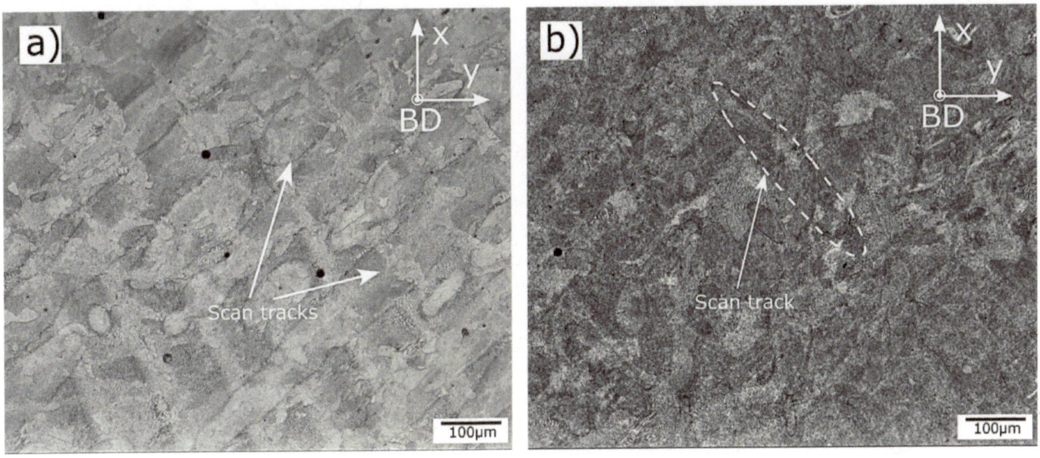

Figure 7. Microstructure of AM samples in the xy-plane (**a**) AM-AP (**b**) AM-HT.

3.2. XRD Analysis and Vickers Hardness Measurements

XRD analysis was performed to confirm the presence of different phases. Figure 8 shows the XRD measurement results for conventional, AM, and AM-HT. For all samples, the primary matrix phase (γ) shows the strongest peaks. Only qualitative phase analysis is performed, as most of the phases give peaks which overlap with the primary matrix, thus making it hard to distinguish them. In the conventional samples, small peaks identified as γ'' are seen close to the γ matrix peaks. Additionally, peaks corresponding to Ti, Nb—carbides are also visible. Even though the microstructure shows the presence of the δ phase, corresponding peaks were not identified in the XRD measurement; this is due to the poor resolution of XRD when the phase fractions are less than 5%.

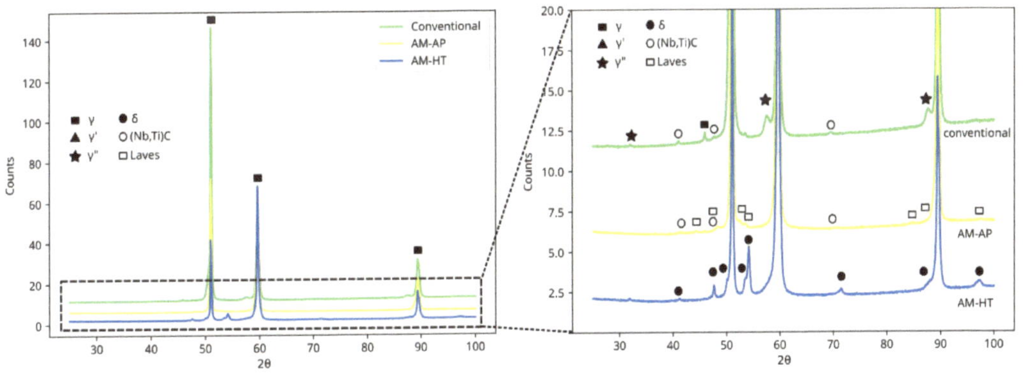

Figure 8. XRD measurement results for Conventional, AM-AP and AM-HT Inconel 718.

For the AM samples in the as-processed condition in addition to the matrix phase, small peaks corresponding to Ti, Nb—carbides and Laves phase are seen. As explained in the previous section, the micro-segregation of Nb and Ti at the interdendritic regions is the reason for the precipitation of these phases.

In the heat-treated samples, the peaks for Laves phase are no longer visible; as during the heat treatment, Laves phase partially dissolves, thus reducing the fraction of phase present. In addition, the dissolution leads to an increase in Nb concentration in the neighbourhood, and the temperature is insufficient for the homogenisation of the elements in the matrix. This leads to the formation of the δ phase in these regions during heat treatment [45]. Small peaks corresponding to the δ phase are visible for the AM-HT sample in Figure 8, confirming the same.

The peaks of γ' and γ'' are difficult to separate from the primary γ matrix, as they are coherent precipitates [12]. However, as the fraction of the precipitates increases, the strength of the material also increases. Thus, hardness measurements can serve as a first estimate for the precipitation of the strengthening phases.

The Vickers hardness measurements of the samples are given in Figure 9. The conventional sample shows a hardness of 380 $\pm$ 13 HV_5, which is in the expected range for the heat-treated wrought alloys [49]. In the case of AM samples, after heat treatment, the sample shows a hardness of 476 $\pm$ 8 HV_5, which is 53% higher than the hardness of the AM-AP sample (310 $\pm$ 39 HV_5). This is expected, as the heat treatment introduces γ' and γ'' strengthening phases, which are absent in the as-processed condition [44]. On the other hand, even after the precipitation of γ' and γ'' phases, the conventional sample shows a hardness value lower than AM-HT samples. The difference in the hardness of AM-HT and conventional samples can be attributed to the fine microstructure induced by the L-PBF process. This is evident when comparing Figures 4 and 6, and is in agreement with data reported by Tanja et al. [50]: where the grain size of L-PBF Inconel 718 is reported to be finer than the forged counterpart by a factor of 10. The higher hardness values of AM-HT would correlate to a higher yield and tensile strength [51]. Thus, when considering hydrogen embrittlement, AM-HT is expected to have a higher degree of HE compared to AM-AP and conventional samples.

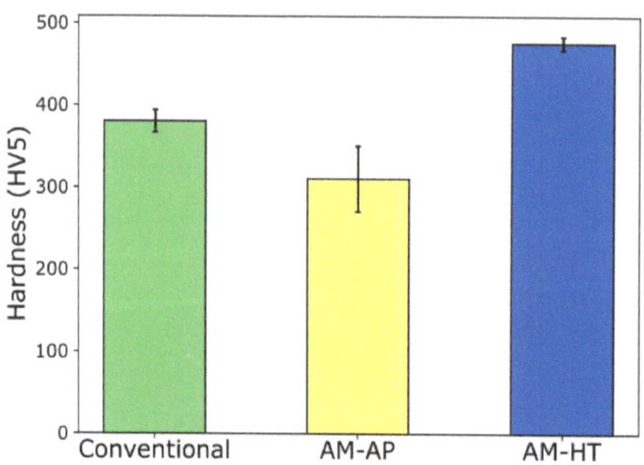

Figure 9. Vickers hardness measurements in yz-plane of conventional, AM-AP and AM-HT Inconel 718.

Hardness measurements are not influenced by the presence of porosity in the L-PBF samples [52]. However, depending on the shape and size of the pores, they can be detrimental to mechanical properties.

Porosity is a defect commonly observed in the components produced from the L-PBF process and is absent in the wrought counterpart, as observed in Figure 10. The AM-AP and AM-HT samples show comparable porosity values of 0.09 $\pm$ 0.02% and 0.05 $\pm$ 0.01%, respectively, making it nearly defect-free (99.9% density) material. Spherical pores signifying gas entrapment were the most commonly seen pore morphology, with the largest pore size being 45 µm. The spherical pores form during the L-PBF process, when the inert gas used to prevent oxidation of powder or the gas already entrapped in the powder feedstock gets trapped in the melt pool during solidification [20]. A few lack of fusion defects (LOF) were also seen; as the name suggests, these defects arise due to insufficient bonding between two layers or scan tracks. Due to their shape, they are detrimental to the mechanical properties as crack initiation can occur at these locations.

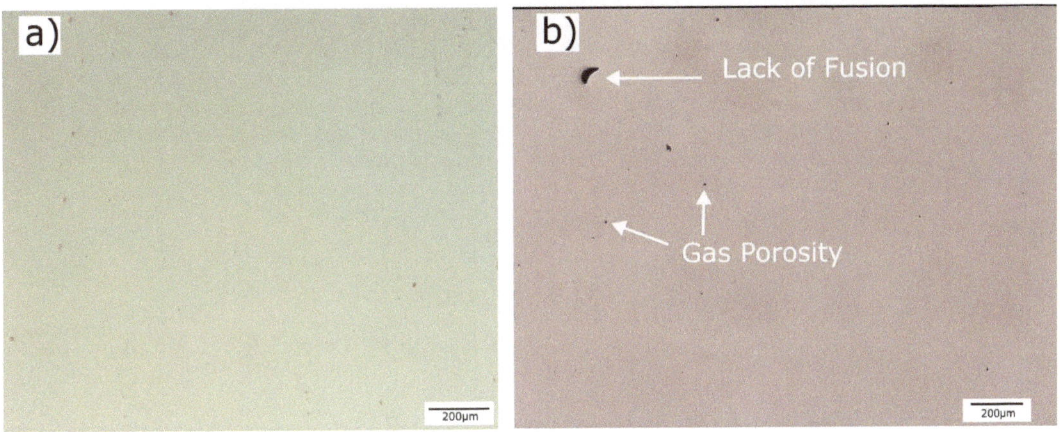

Figure 10. Representative images of porosity for (**a**) Conventional and (**b**) AM Inconel 718.

3.3. In-Situ Gaseous Hydrogen Slow Strain Rate Testing (SSRT)

The results of the SSRT test are summarised in Table 4 and representative plots for each testing condition are given in Figure 11. As can be seen, the major difference occurs with the AM heat-treated samples where the double ageing with longer duration is employed. With the precipitation of strengthening phases, an increase of 62% in the yield strength (Y.S.) and a 57% increase in the ultimate tensile strength (UTS) was seen for the AM-HT sample. As accurate quantification of the strengthening phases is not performed in this study, information from the literature is used as reference.

Table 4. In-situ slow strain rate testing (SSRT) results.

Sample	Environment *	Y.S. (MPa)	UTS (MPa)	Elongation to Failure (%)	RA (%)
Conventional	Nitrogen	1069 ± 4	1474 ± 28	12.7 ± 1.4	30.6 ± 1.4
	Hydrogen	1059 ± 34	1308 ± 01	5.8 ± 0.1	17.2 ± 2.1
AM-AP	Nitrogen	898 ± 11	1129 ± 11	7.5 ± 0.1	24.0 ± 1.3
	Hydrogen	897 ± 12	1153 ± 05	5.4 ± 0.1	19.5 ± 0.6
AM-HT	Nitrogen	1462 ± 08	1776 ± 07	4.5 ± 0.3	11.7 ± 1.8
	Hydrogen	1409 ± 10	1610 ± 03	2.8 ± 0.2	10.8 ± 0.3

* Nitrogen and hydrogen testing were performed in in-situ condition with 150 bar and 25 °C. For hydrogen testing, an additional 48 h charging was performed at 25 °C.

Embrittlement is observed in both conventional and AM samples when hydrogen was introduced. However, the degree of hydrogen embrittlement varies. A general observation for hydrogen samples which was not found in the literature was the drop in load after reaching particular stress. In a uniaxial tensile test, a load drop before failure signifies a reduction in load-bearing cross-sectional area or softening. As no significant load drop in the nitrogen test condition is observed it is unlikely that softening occurs. This can be further confirmed by looking at the fracture surface as occurs in a later section.

Hydrogen embrittlement index can serve as a quantitative comparison between conventional and AM samples. The hydrogen embrittlement index (HEI) of the samples are provided in Figure 12 with higher values signifying a higher degree of embrittlement. The AM samples show lower degree of embrittlement than the conventional samples. In the case of AM-AP samples, the difference could arise due to the absence of strengthening phases γ' and γ''. Rezende et al. [53] studied the influence of γ'' precipitates on the HE behaviour and reported that its presence aggravated hydrogen embrittlement. Similar observations were also seen by Hesketh et al. [33], where the lower degree of HE was attributed to the lower tensile strength of AM-AP samples. In general metals having high tensile strength show a higher degree of hydrogen embrittlement [36]. Surprisingly in this study, the AM-HT samples, which demonstrate the highest yield strength, show the least embrittlement compared to AM-AP and conventional samples, thus deviating from the earlier mentioned trend. This behaviour could arise due to the unique anisotropic microstructure inherent to the L-PBF process, which could affect how hydrogen interacts with the material. To better understand this, the fracture surface is looked into to help identify the HE mechanisms.

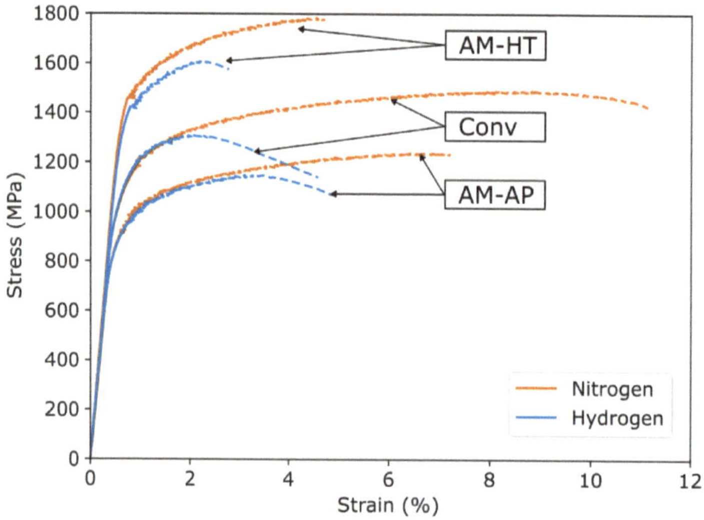

Figure 11. Representative SSRT curves for conventional, AM-AP and AM-HT Inconel 718.

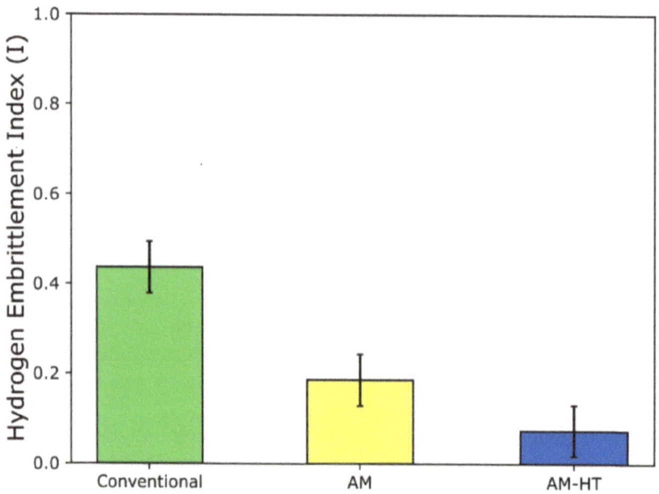

Figure 12. HE indexes of conventional, AM-AP and AM-HT Inconel 718 samples.

3.4. Fractography

The fracture surfaces of the conventional and AM samples tested in ambient conditions are given in Figures 13 and 14 respectively. Conventional samples tested in a nitrogen environment (Figure 13a) show features of a typical ductile failure, i.e., failure by micro-void coalescence (Figure 13b). The void size was determined to be ~4.5 μm.

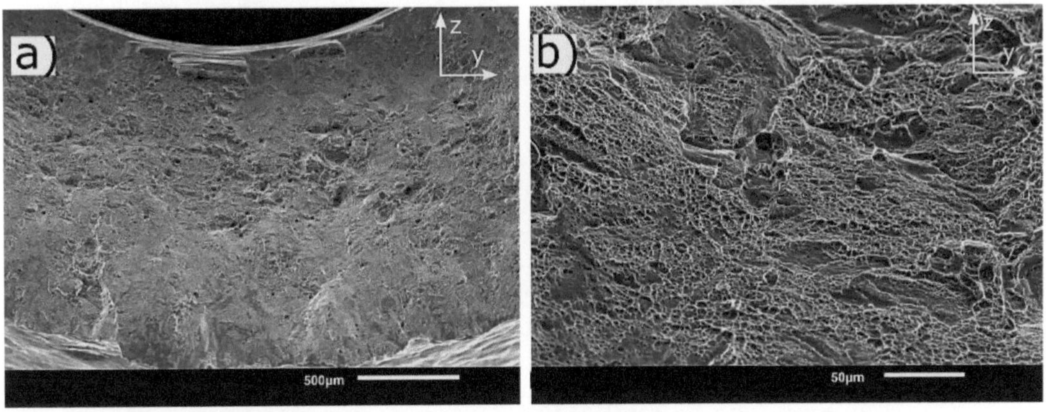

Figure 13. (**a**) Zoomed-out and (**b**) zoomed-in view showing characteristic ductile fracture with micro-voids in conventional Inconel 718 tested in nitrogen.

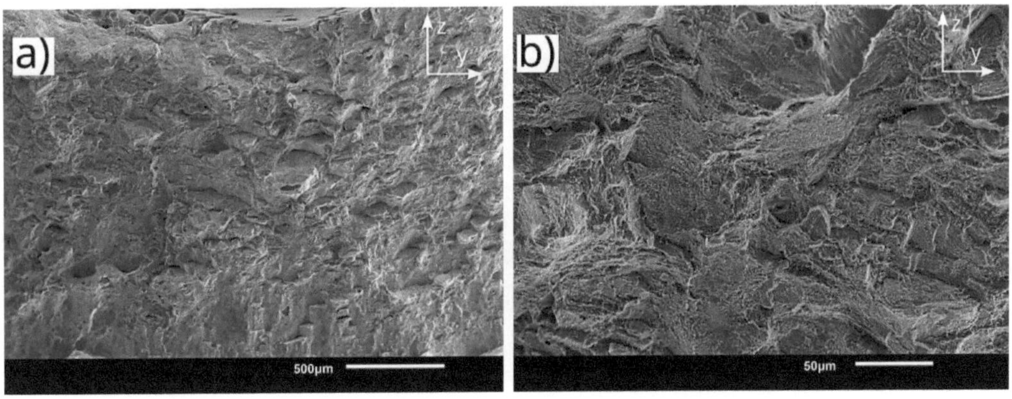

Figure 14. Fracture surface of AM-AP tested in nitrogen at 150 bar. (**a**) Low magnification. (**b**) High magnification.

In the case of AM samples, both AM-AP and AM-HT show a ductile mode of failure in a nitrogen environment; however, the micro-voids appear to be ~65% smaller than the conventional samples. The smaller dimple size (~1.5 μm) is attributed to the fine microstructure produced from the L-PBF process [45,54]. As no significant variation is noted in the fracture surface of AM-AP and AM-HT samples, only one representative surface is shown here (Figure 14).

The conventional samples after hydrogen gaseous charging show severe embrittlement on the fracture surface, as seen in Figure 15. Additionally, it is noted that fracture initiates from the inner surface of the sample. Three distinct regions can be observed on the surface (Figure 15): (1) brittle intergranular region, (2) quasi-cleavage region and (3) ductile region. Severe cracking (marked by arrows) on the fracture surface is also seen.

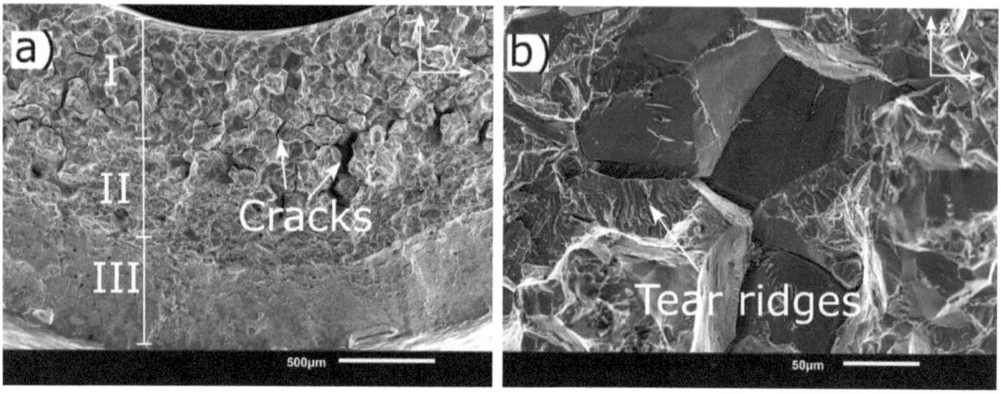

Figure 15. Three distinct regions on the fracture surface of conventional samples tested in hydrogen in-situ condition. (**a**) low magnification (**b**) high magnification at region of intergranular failure.

The brittle region in the in-situ sample consisted of features representative of the grain structure. Tear ridges signifying plasticity are also visible in this region. Several researchers have reported fracture surfaces similar to this after electrochemical charging [14,55]. Zhang et al. [55] reported that though the features show an intergranular mode of failure, the crack initiates and propagates through dislocation slip bands and hence is actually transgranular in nature. In this case, the proposition by Zhang et al. [55] that failure initiates at the slip bands requires further analysis, as it is unlikely that hydrogen has sufficient time for diffusing within the grain.

To identify the dominant mechanism, the intergranular region on the fracture surface is analysed. On a macro scale, the brittle features point towards hydrogen-enhanced decohesion (HEDE) mechanism [17]. However, on taking a closer look at the facets, dislocation slip lines signifying plastic deformation are visible (Figure 16). This indicates a possibility for hydrogen-enhanced localised plasticity (HELP) or adsorption-induced dislocation emission (AIDE) mechanism [17]. AIDE is considered here as the hydrogen is present on the surfaces of the crack, which can be adsorbed leading to dislocation emission from the crack tip.

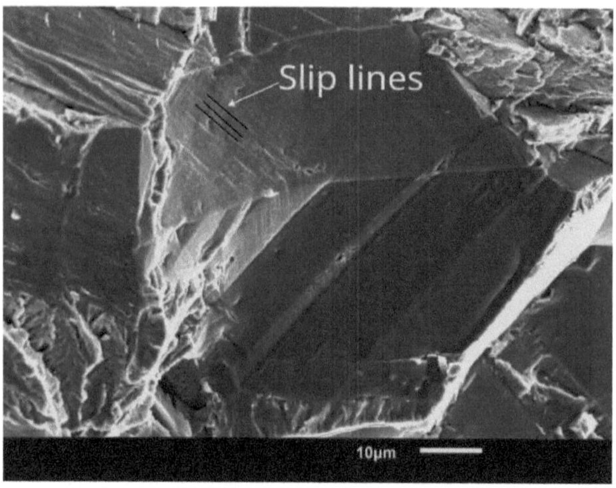

Figure 16. Slip lines along the brittle facets of in-situ hydrogen embrittled samples.

Another observation that needs to be accounted for is the load drop in the SSRT curves (Figure 11). It was found that the drop in load occurred only after a particular stress level is reached. As explained earlier this means a reduction in load-bearing area; here, it signifies the beginning of decohesion. The strain at this point was determined to be 2%. As Inconel 718 has a low stacking fault energy, it is found to show planar deformation [56]. The slip lines visible on the facets could be a consequence of this 2% strain. Further, it could also be possible that the decohesion occurred where the slip lines intersected with the grain boundaries. A cross-section of the fracture surface was analysed to understand it better.

Figure 17 shows the cross-section of fracture at the brittle region; both intergranular and transgranular modes of failure are observed. A large number of dislocation slip lines are also visible; the transgranular failure could have occurred along these slip lines. The tear ridges seen in Figure 15 confirm this. A closer look with a higher resolution electron microscope could help better understand why there was a transition from intergranular to transgranular mode of failure. Decohesion along grain boundaries ahead of the crack is also seen. It is unlikely that hydrogen diffuses far ahead for the decohesion to occur during the test. This decohesion could be a consequence of another crack front in the plane. It is likely that the mode of failure is an interplay between HEDE, HELP and AIDE mechanisms.

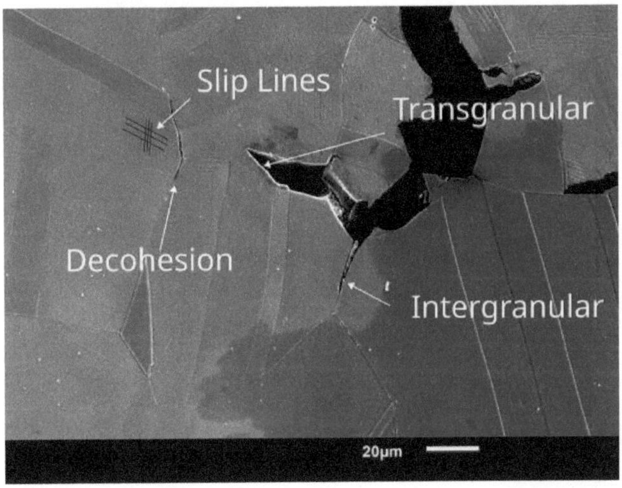

Figure 17. Cross-sectional image of fracture surface showing different modes of failure in conventional Inconel 718.

Similar to the conventional samples, the hydrogen AM samples also show a severely embrittled region (Figures 18 and 19). A distinct difference observed here is that the HE region is directional for all AM samples (both AM-AP and AM-HT), unlike the conventional samples, which show homogeneous embrittlement throughout the fracture plane. A representative macroscopic fracture surface from an AM sample and conventional sample is given in Figure 20. The observed difference could arise due to the microstructural anisotropy in the L-PBF samples. Features representative of elongated grains are visible in the HE region for both AM-AP and AM-HT samples. Further, these HE regions appear to be oriented along the build direction, as can be deduced from these elongated features. The lower degree of HE in AM samples compared to the conventional samples could be a consequence of this preferential anisotropy; as only a small region is embrittled and the remainder of the sample retains its ductility. To better understand this, the microstructure in the fracture plane was analysed.

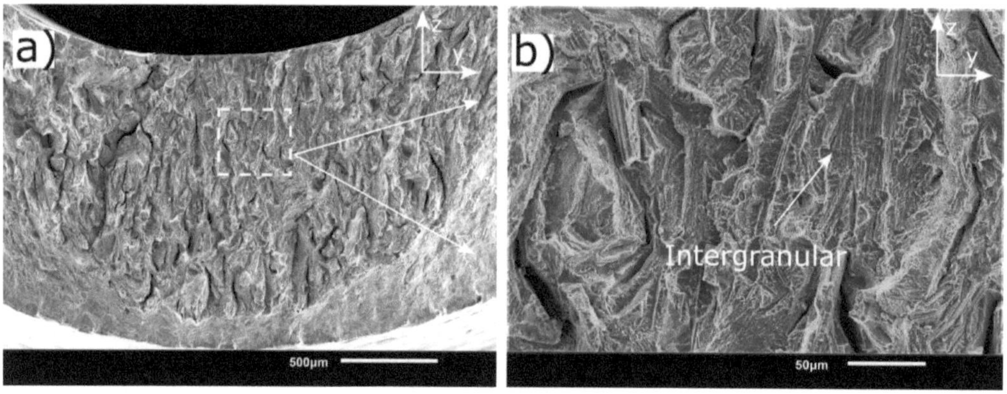

Figure 18. Hydrogen embrittled fracture region in AM-AP sample. (**a**) Low magnification. (**b**) High magnification of marked region.

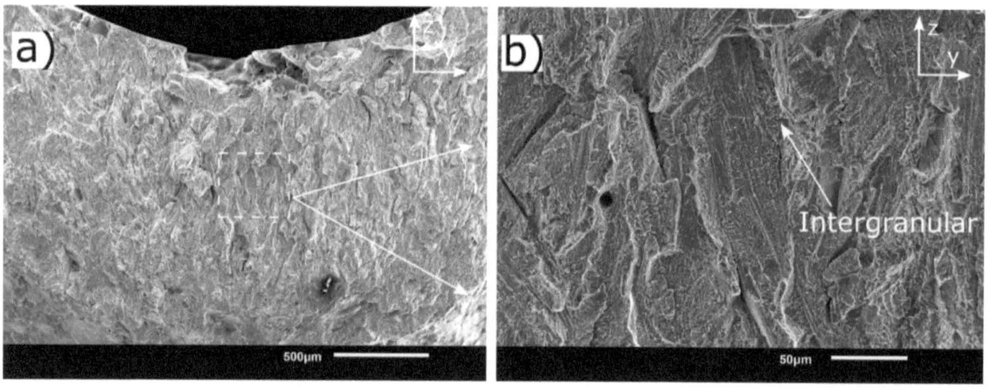

Figure 19. Hydrogen embrittled fraction region in AM-HT sample. (**a**) Low magnification. (**b**) High magnification of marked region.

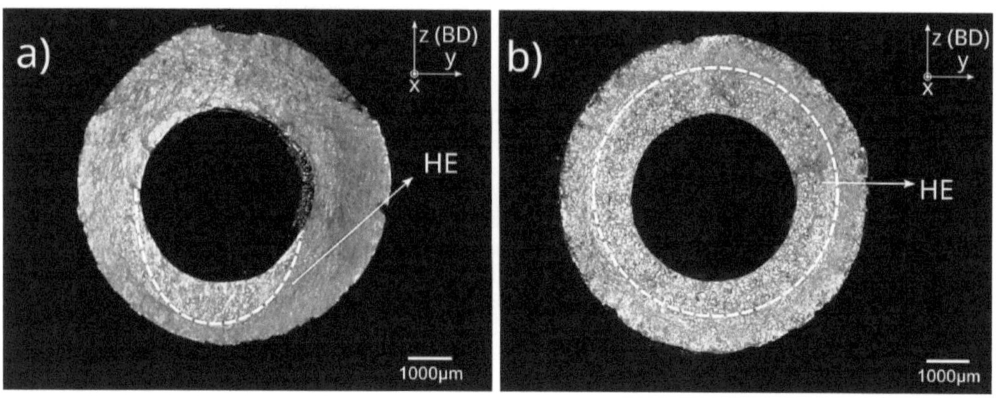

Figure 20. (**a**) Anisotropy in the HE region of AM samples. (**b**) Homogeneous embrittled region in conventional sample. X-axis is the tensile loading direction and z-axis is the build direction.

In the conventional sample, in all planes (xy, xz and yz-planes) a uniform equiaxed microstructure is observed. Microstructure along the fracture plane (yz-plane) is shown in Figure 21a). This means that after crack initiation, its propagation occurs the same way in all the directions of the fracture plane (yz-plane). Even if multiple cracks initiate at the inner surface as seen in the case of conventional samples, the energy required for crack propagation remains the same.

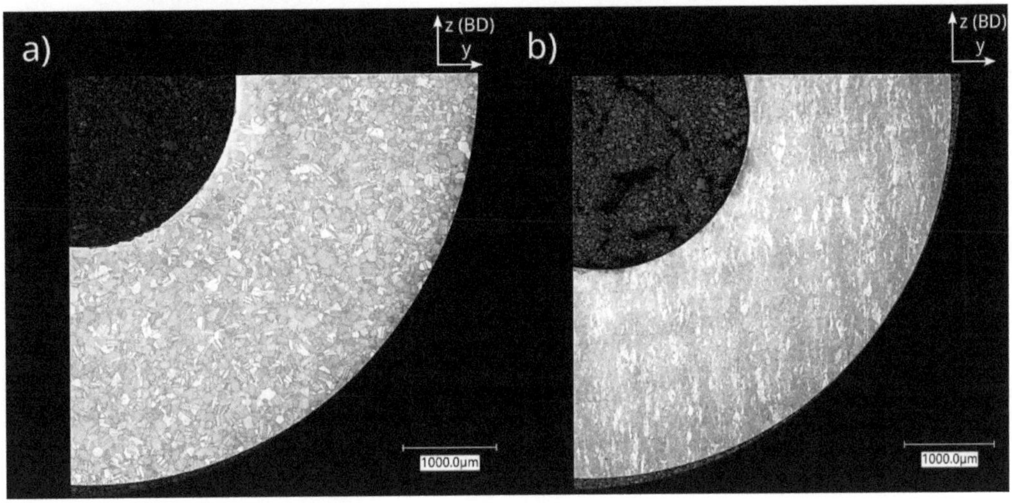

Figure 21. Microstructure in the fracture plane for (**a**) Conventional Inconel 718 showing equiaxed grains, (**b**) AM-HT showing elongated grains. As AM-HT clearly shows elongated grains without meltpools, it is used as the representative image for all AM samples.

As seen earlier, it is clear that the AM samples show different microstructures along different planes, with both AM-AP and AM-HT showing similar features. The microstructure along the fracture plane (yz-plane), as seen in Figure 21b), shows elongated grains oriented along the L-PBF build direction. A similar microstructure is also seen in the xz-plane; however, in the xy-plane the microstructure is different. As explained in microstructure section, it is because xy-plane is the laser scanning plane. A schematic representation of this difference in the fracture plane is given in Figure 22.

Crack propagation occurs in material when there is sufficient energy for creating two new surfaces [57]; as it is a brittle mode of failure energy for plastic deformation is not considered here. In a uniaxial tensile test, this energy is provided by the tensile load. If Es is the energy required to create two new surfaces, and assuming that the elongated grains are of the similar grain boundary area. The energy required for the crack to travel across a grain boundary surface would be the same in both region A and B (Figure 22). This is because the area of the two new surfaces created by the crack would be the same. However, the distance the crack travels by expending this energy is much larger in the z-direction (build direction) than in the y-direction; because the longer axis of elongated grains is oriented in the crack propagation direction as shown in Figure 22c. This is confirmed from the fracture surfaces, where along the y-direction the embrittled region is found to be ≈175 µm wide, while along the z-direction the embrittled region is an order of magnitude higher (≈1.5 mm). A more in-depth analysis is required to understand why there is an order of magnitude difference in the embrittled region.

Anisotropy in the vertically (tensile axis along the z-axis) and horizontally built samples (tensile axis along the x or y-axis) are commonly reported in the literature [20]. Hesketh et al. [33] studied the effect of orientation on hydrogen embrittlement of L-PBF Inconel 718 by in-situ electrochemical charging, and reported that the horizontal cylindrical samples exhibited higher degree of embrittlement compared to vertically built ones. The differences in the experimental conditions in the two studies makes it difficult to perform a quantitative analysis. Nevertheless, the trend in anisotropy is established. A similar study was done by Li et al. [31], where the HE behaviour of AM-produced Inconel 718 on dog bone specimens with two different horizontal orientations was looked into, keeping the long axis parallel to the tensile axis. It was reported that the sample with a microstructure similar to the Figure 22 c showed a higher degree of HE. This confirms the anisotropy in flat specimens built in two different horizontal directions. In a ductile mode of failure, it is difficult to observe features on the fracture surface which could be a result of this anisotropy within a horizontally built sample. This further shows, the need for careful design considerations when using L-PBF components.

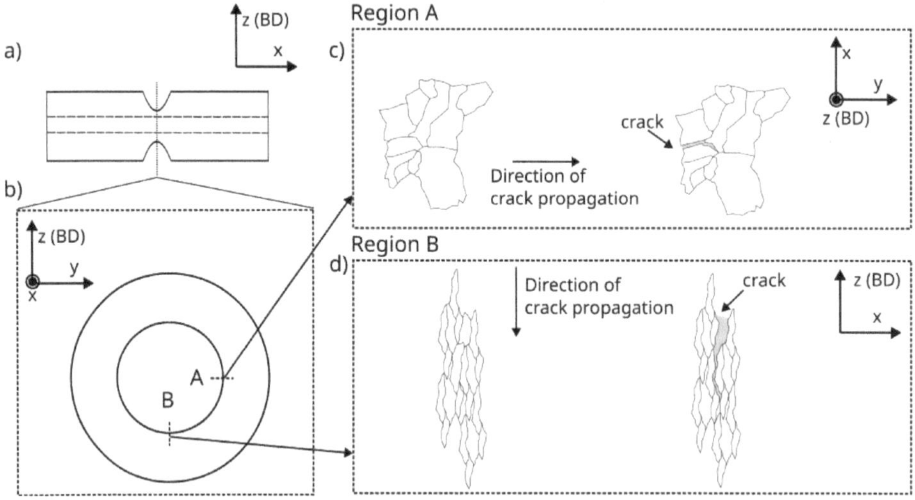

Figure 22. (**a**) Schematic of L-PBF sample showing area of interest (notch). (**b**) Cross-sectional view of the notch. (**c**) Crack propagation in xy-plane. (**d**) Crack propagation in xz-plane. z-axis is the build direction (BD) and x-axis is the tensile axis. The microstructure schematic is not to scale.

Among the AM-AP and AM-HT samples (Figures 18 and 19, respectively), the former shows prominent brittle features and cracks on the fracture surface. This is in line with the observation of higher HEI for the AM-AP compared to AM-HT in the previous section. At high magnification (Figure 23), continuous dendritic structures are visible in the AM-AP samples inside the grain-like features; whereas, in the AM-HT samples the dendritic structures appear discontinuous. Li et al. [31] reported that in the as-processed state, the hydrogen-assisted cracking originated between γ matrix and Laves phase. Hence, failure may occur more easily along the dendritic structures in the as-processed state, thus leading to a higher degree of embrittlement. This is also supported by the observation of Lin et al. [58], where it was reported that the cellular structures in L-PBF produced 316 L stainless steel enhanced the diffusion of hydrogen and induced defects. This could be the reason for the features to be less prominent in the heat-treated condition as partial dissolution of Laves phase occurs [48], as seen in the microstructure section.

Li et al. [31] and Lee et al. [30] reported that cellular dendritic structures in the AM as-processed state did not influence HE. The contrasting results could be a consequence of testing conditions, sample geometry and orientations. As seen here, the anisotropy led to directional crack growth in hollow cylindrical samples. These features would not be clearly evident in flat dog-bone specimens. Additionally in these studies, SSRT tests were performed after electrochemical pre-charging which means the concentration of hydrogen remained similar in both as-processed and heat-treated conditions. As the yield strength of the as-processed samples are low, for the same hydrogen concentration as heat-treated samples, the degree of embrittlement would be low. However, in this study, there is a continuous supply of hydrogen at the crack surfaces leading to a higher degree of embrittlement.

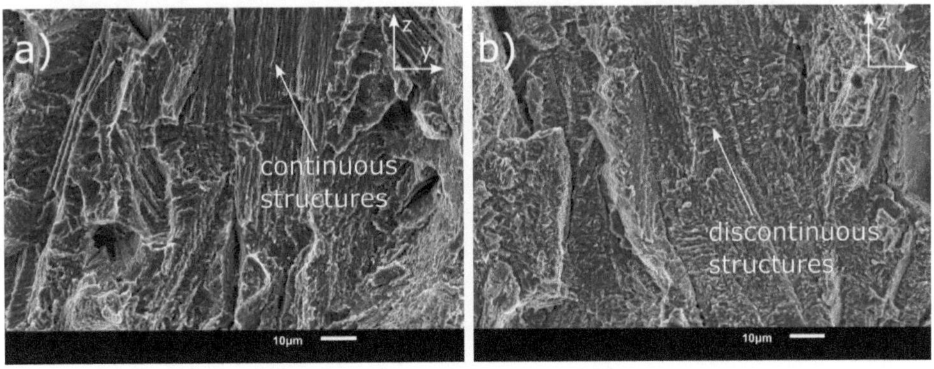

Figure 23. High magnification image of grain like features on the fracture surface of AM samples. (**a**) AM-AP showing continuous structures. (**b**) AM-HT showing discontinuous structures.

Another aspect that needs to be considered is the experiment duration. As the crosshead speed is kept constant, the duration of experiment for each set of samples is different. This is more pronounced for samples with different yield strengths, as they begin to strain plastically at different times. To understand this better, consider an elongation of 1 mm; the time taken for this elongation is same for all samples, as it is displacement-controlled. However, as elongation is different for each sample, the duration of the experiment is different for each sample set. This would lead to a difference in the hydrogen uptake and diffusion during the experiment (although the same 48 h pre-charge time was applied for all samples). The elongation of the AM-AP samples at failure was found to be $\approx$5.4%, whereas for AM-HT it was only 2.8%. This means after yielding, HE in AM-AP sample occurs in a longer duration than the AM-HT samples. This could be the major reason why higher degree of embrittlement is observed. Tarzimoghadam [59] had similar observations, where it was reported that the annealed specimens showed a $\sim$70% drop in elongation even though the intermetallic precipitates were absent and the yield strength was only 400 MPa.

Although porosities were observed on the fracture surface of AM samples, they did not appear to influence HE, which is due to the negligible amount of porosity in L-PBF samples. Hesketh et al. [33] also reported similar findings. In theory, porosity would be detrimental as it would reduce the cross-sectional area and can act as a stress concentrator, aggravating hydrogen embrittlement.

4. Conclusions

This study focused on the in-situ hydrogen gas embrittlement behaviour of 3D printed Inconel 718, produced from recycled powder as compared to the conventional wrought counterpart. Based on the study, the following conclusions can be made:

- The L-PBF Inconel 718 produced from recycled powder shows severe embrittlement in the presence of in-situ gaseous (150 bar) hydrogen. However, in the as-processed state, the degree of embrittlement (HEI) is 50% lower compared to the conventional Inconel 718.
- L-PBF Inconel 718 was found to have coarse-grained microstructure with columnar grains elongated along the building direction (BD). Microstructural anisotropy with respect to BD was observed in both as-processed and after post-processing heat

treatment states. Such anisotropy has led to a narrow hydrogen embrittlement zone along the build direction.

- Despite the higher yield strength, AM-HT Inconel 718 showed the least degree of hydrogen embrittlement (64% lower than conventional Inconel 718). The better resistance compared to the conventional Inconel 718 could be attributed to the AM-induced microstructural anisotropy. As hydrogen embrittlement occurs only in a narrow region along the build direction, the remainder of the material retains its ductility, unlike the conventional sample where a uniform and wide HE region is observed.
- The fracture surface of AM Inconel 718 was found to be similar to that of the conventional samples. It can, thus, be expected that the same HE mechanisms is occurring in both conventional and AM samples. For the conventional samples, brittle cleavage facets and presence of dislocation slip lines signifying plasticity were found on the fracture surface. It was concluded that the HE mechanism includes an interplay between HEDE, HELP and AIDE.

To delineate the effect of powder on hydrogen embrittlement, it is recommended to test L-PBF components produced from the conventional Inconel 718 powder. Additionally, in this study it was seen that the unique anisotropy in microstructure is the cause for lower hydrogen embrittlement index of AM-HT samples. This suggests that, with careful design considerations, for L-PBF produced components, properties can be selectively tailored at regions exposed to hydrogen to minimise hydrogen embrittlement.

Author Contributions: Conceptualization, N.K.M., A.G., M.V., T.R., S.P.S. and V.P.; investigation, N.K.M.; writing—original draft preparation, N.K.M.; writing—review and editing, V.P., T.R., S.P.S. and M.V.; supervision, V.P. All authors have read and agreed to the published version of the manuscript.

Funding: This research received no external funding.

Institutional Review Board Statement: Not applicable.

Informed Consent Statement: Not applicable.

Data Availability Statement: The data presented in this study are available in the article.

Conflicts of Interest: The authors declare no conflict of interest.

References

1. Gibson, I.; Rosen, D.; Stucker, B. *Additive Manufacturing Technologies 3D Printing, Rapid Prototyping, and Direct Digital Manufacturing*, 2nd ed.; Springer: New York, NY, USA, 2010.
2. Javaid, M.; Haleem, A.; Singh, R.P.; Suman, R.; Rab, S. Role of additive manufacturing applications towards environmental sustainability. *Adv. Ind. Eng. Polym. Res.* **2021**, *4*, 312–322. [CrossRef]
3. Bhowmik, S.; McWilliams, B.A.; Knezevic, M. Effect of powder reuse on tensile, compressive, and creep strength of Inconel 718 fabricated via laser powder bed fusion. *Mater. Charact.* **2022**, *190*, 112023. [CrossRef]
4. Bjørge, O.K. Mechanical Properties of 316L Stainless Steel Made by Selective Laser Melting Using Powder Produced from Recycled Scrap Material by Vacuum Induction Gas Atomizing. Bachelor's Thesis, Universitetet i Stavanger, Stavanger, Norway, 2021.
5. Fullenwider, B.; Kiani, P.; Schoenung, J.M.; Ma, K. *From Recycled Machining Waste to Useful Powders for Metal Additive Manufacturing*; Springer: Cham, Switzerland, 2019; pp. 3–7.
6. Yi, F.; Zhou, Q.; Wang, C.; Yan, Z.; Liu, B. Effect of powder reuse on powder characteristics and properties of Inconel 718 parts produced by selective laser melting. *J. Mater. Res. Technol.* **2021**, *13*, 524–533. [CrossRef]
7. Ardila, L.C.; Garciandia, F.; González-Díaz, J.B.; Álvarez, P.; Echeverria, A.; Petite, M.M.; Deffley, R.; Ochoa, J. Effect of IN718 Recycled Powder Reuse on Properties of Parts Manufactured by Means of Selective Laser Melting. *Phys. Procedia* **2014**, *56*, 99–107. [CrossRef]
8. Gruber, K.; Smolina, I.; Kasprowicz, M.; Kurzynowski, T. Evaluation of Inconel 718 Metallic Powder to Optimize the Reuse of Powder and to Improve the Performance and Sustainability of the Laser Powder Bed Fusion (LPBF) Process. *Materials* **2021**, *14*, 1538. [CrossRef]
9. Sutton, A.T.; Kriewall, C.S.; Leu, M.C.; Newkirk, J.W. Characterization of heat-affected powder generated during the selective laser melting of 304L stainless steel powder. In Proceedings of the Solid Freeform Conference, Austin, TX, USA, 7–9 August 2017; pp. 7–9.

10. Slotwinski, J.A.; Garboczi, E.J.; Stutzman, P.E.; Ferraris, C.F.; Watson, S.S.; Peltz, M.A. Characterization of Metal Powders Used for Additive Manufacturing. *J. Res. Natl. Inst. Stand. Technol.* **2014**, *119*, 460. [CrossRef]
11. Sutton, A.T.; Kriewall, C.S.; Karnati, S.; Leu, M.C.; Newkirk, J.W.; Everhart, W.; Brown, B. Evolution of AISI 304L stainless steel part properties due to powder recycling in laser powder-bed fusion. *Addit. Manuf.* **2020**, *36*, 101439. [CrossRef]
12. Reed, R.C. *Superalloys—Fundamentals and Applications*; Cambridge University Press: Cambridge, UK, 2006.
13. Soller, S.; Barata, A.; Beyer, S.; Dahlhaus, A.; Guichard, D.; Humbert, E.; Kretschmer, J.; Zeiss, W. Selective laser melting (SLM) of Inconel 718 and stainless steel injectors for liquid rocket engines. In Proceedings of the Space Propulsion Conference 2016, Roma, Italy, 2–6 May 2016.
14. Zhang, Z.; Obasi, G.; Morana, R.; Preuss, M. Hydrogen assisted crack initiation and propagation in a nickel-based superalloy. *Acta Mater.* **2016**, *113*, 272–283. [CrossRef]
15. Liu, L.; Tanaka, K.; Hirose, A.; Kobayashi, K. Effects of precipitation phases on the hydrogen embrittlement sensitivity of Inconel 718. *Sci. Technol. Adv. Mater.* **2002**, *3*, 335–344. [CrossRef]
16. Ehrlin, N.; Bjerkén, C.; Fisk, M. Cathodic hydrogen charging of Inconel 718. *AIMS Mater. Sci.* **2016**, *3*, 1350–1364. [CrossRef]
17. Lynch, S. Hydrogen embrittlement phenomena and mechanisms. *Corros. Rev.* **2012**, *30*, 105–123. [CrossRef]
18. *ASTM F2078-22 Red*; Standard Terminology Relating to Hydrogen Embrittlement Testing. ASTM International: West Conshohocken, PA, USA, 2016.
19. Popovich, V.A.; Borisov, E.V.; Popovich, A.A.; Sufiiarov, V.S.; Masaylo, D.V.; Alzina, L. Functionally graded Inconel 718 processed by additive manufacturing: Crystallographic texture, anisotropy of microstructure and mechanical properties. *Mater. Des.* **2017**, *114*, 441–449. [CrossRef]
20. DebRoy, T.; Wei, H.L.; Zuback, J.S.; Mukherjee, T.; Elmer, J.W.; Milewski, J.O.; Beese, A.M.; Wilson-Heid, A.; De, A.; Zhang, W. Additive manufacturing of metallic components—Process, structure and properties. *Prog. Mater. Sci.* **2018**, *92*, 112–224. [CrossRef]
21. Brenne, F.; Taube, A.; Pröbstle, M.; Neumeier, S.; Schwarze, D.; Schaper, M.; Niendorf, T. Microstructural design of Ni-base alloys for high-temperature applications: Impact of heat treatment on microstructure and mechanical properties after selective laser melting. *Prog. Addit. Manuf.* **2016**, *1*, 141–151. [CrossRef]
22. Schneider, J.; Lund, B.; Fullen, M. Effect of heat treatment variations on the mechanical properties of Inconel 718 selective laser melted specimens. *Addit. Manuf.* **2018**, *21*, 248–254. [CrossRef]
23. Gallmeyer, T.G.; Moorthy, S.; Kappes, B.B.; Mills, M.J.; Amin-Ahmadi, B.; Stebner, A.P. Knowledge of process-structure-property relationships to engineer better heat treatments for laser powder bed fusion additive manufactured Inconel 718. *Addit. Manuf.* **2020**, *31*, 100977. [CrossRef]
24. Liu, P.; Hu, J.; Sun, S.; Feng, K.; Zhang, Y.; Cao, M. Microstructural evolution and phase transformation of Inconel 718 alloys fabricated by selective laser melting under different heat treatment. *J. Manuf. Process.* **2019**, *39*, 226–232. [CrossRef]
25. Merson, E.; Myagkikh, P.; Poluyanov, V.; Dorogov, M.; Merson, D.; Vinogradov, A. The fundamental difference between cleavage and hydrogen-assisted quasi-cleavage in ferritic materials revealed by multiscale quantitative fractographic and side surface characterization. *Mater. Sci. Eng. A* **2021**, *824*, 141826. [CrossRef]
26. Lynch, S.P. Progress towards understanding mechanisms of hydrogen embrittlement and stress corrosion cracking. In Proceedings of the CORROSION 2007, Nashville, TN, USA, 11–15 March 2007.
27. Nagumo, M. *Fundamentals of Hydrogen Embrittlement*; Springer: Singapore, 2016; Volume 921.
28. Hicks, P.D.; Altstetter, C.J. Hydrogen-enhanced cracking of superalloys. *Metall Trans. A* **1992**, *23*, 237–249. [CrossRef]
29. Fournier, L.; Delafosse, D.; Magnin, T. Cathodic hydrogen embrittlement in alloy 718. *Mater. Sci. Eng. A* **1999**, *269*, 111–119. [CrossRef]
30. Lee, D.-H.; Zhao, Y.; Lee, S.Y.; Ponge, D.; Jägle, E.A. Hydrogen-assisted failure in Inconel 718 fabricated by laser powder bed fusion: The role of solidification substructure in the embrittlement. *Scr. Mater.* **2022**, *207*, 114308. [CrossRef]
31. Li, X.; Li, Q.; Wang, T.; Zhang, J. Hydrogen-assisted failure of laser melting additive manufactured IN718 superalloy. *Corros. Sci.* **2019**, *160*, 108171. [CrossRef]
32. Aiello, F.; Beghini, M.; Bertini, L.; Macoretta, G.; Monelli, B.D.; Valentini, R. Hydrogen diffusivity and tensile properties degradation in SLMed Inconel 718. *IOP Conf. Ser. Mater. Sci. Eng.* **2022**, *1214*, 012002. [CrossRef]
33. Hesketh, J.; McClelland, N.; Zhang, Y.; Green, C.; Turnbull, A. Influence of additive manufacturing by laser powder bed fusion on the susceptibility of Alloy 718 to hydrogen embrittlement. *Corros. Eng. Sci. Technol.* **2021**, *56*, 565–574. [CrossRef]
34. F3nice. Available online: https://f3nice.com/ (accessed on 1 October 2022).
35. *ASTM F3055-14a*; Standard Specification for Additive Manufacturing Nickel Alloy (UNS N07718) with Powder Bed Fusion. ASTM International: West Conshohocken, PA, USA, 2021.
36. Nagumo, M. Characteristic Features of Deformation and Fracture in Hydrogen Embrittlement. In *Fundamentals of Hydrogen Embrittlement*; Springer: Singapore, 2016; pp. 137–165.
37. *AMS5663N*; Nickel Alloy, Corrosion and Heat-Resistant, Bars, Forgings, and Rings 52.5Ni–19Cr–3.0Mo–5.1Cb (Nb)–0.90Ti–0.50Al–18Fe Consumable Electrode or Vacuum Induction Melted 1775 °F (968 °C) Solution and Precipitation Heat Treated. SAE International: Warrendale, PA, USA, 2016.
38. VDM Metals GmbH. Available online: https://www.vdm-metals.com/en/ (accessed on 1 October 2022).
39. *ASTM B637-18*; Book of Standards vol. 02.04. ASTM International: West Conshohocken, PA, USA, 2018.

40. Boot, T.; Riemslag, T.A.C.; Reinton, E.T.E.; Liu, P.; Walters, C.L.; Popovich, V. In-Situ Hollow Sample Setup Design for Mechanical Characterisation of Gaseous Hydrogen Embrittlement of Pipeline Steels and Welds. *Metals* **2021**, *11*, 1242. [CrossRef]
41. Ganji, D.K.; Rajyalakshmi, G. Influence of Alloying Compositions on the Properties of Nickel-Based Superalloys: A Review. In *Lecture Notes in Mechanical Engineering*; Springer: Singapore, 2020; pp. 537–555.
42. Cai, D.; Zhang, W.; Nie, P.; Liu, W.; Yao, M. Dissolution kinetics of delta phase and its influence on the notch sensitivity of Inconel 718. *Mater. Charact.* **2007**, *58*, 220–225. [CrossRef]
43. Li, X.; Shi, J.J.; Wang, C.H.; Cao, G.H.; Russell, A.M.; Zhou, Z.J.; Li, C.P.; Chen, G.F. Effect of heat treatment on microstructure evolution of Inconel 718 alloy fabricated by selective laser melting. *J. Alloy. Compd.* **2018**, *764*, 639–649. [CrossRef]
44. Chlebus, E.; Gruber, K.; Kuźnicka, B.; Kurzac, J.; Kurzynowski, T. Effect of heat treatment on the microstructure and mechanical properties of Inconel 718 processed by selective laser melting. *Mater. Sci. Eng. A* **2015**, *639*, 647–655. [CrossRef]
45. Zhang, D.; Niu, W.; Cao, X.; Liu, Z. Effect of standard heat treatment on the microstructure and mechanical properties of selective laser melting manufactured Inconel 718 superalloy. *Mater. Sci. Eng. A* **2015**, *644*, 32–40. [CrossRef]
46. Sabelkin, V.P.; Cobb, G.R.; Shelton, T.E.; Hartsfield, M.N.; Newell, D.J.; O'Hara, R.P.; Kemnitz, R.A. Mitigation of anisotropic fatigue in nickel alloy 718 manufactured via selective laser melting. *Mater. Des.* **2019**, *182*, 108095. [CrossRef]
47. Wang, L.Y.; Zhou, Z.J.; Li, C.P.; Chen, G.F.; Zhang, G.P. Comparative investigation of small punch creep resistance of Inconel 718 fabricated by selective laser melting. *Mater. Sci. Eng. A* **2019**, *745*, 31–38. [CrossRef]
48. Popovich, V.A.; Borisov, E.V.; Popovich, A.A.; Sufiiarov, V.S.; Masaylo, D.V.; Alzina, L. Impact of heat treatment on mechanical behaviour of Inconel 718 processed with tailored microstructure by selective laser melting. *Mater. Des.* **2017**, *131*, 12–22. [CrossRef]
49. Hosseini, E.; Popovich, V.A. A review of mechanical properties of additively manufactured Inconel 718. *Addit. Manuf.* **2019**, *30*, 100877. [CrossRef]
50. Trosch, T.; Strößner, J.; Völkl, R.; Glatzel, U. Microstructure and mechanical properties of selective laser melted Inconel 718 compared to forging and casting. *Mater. Lett.* **2016**, *164*, 428–431. [CrossRef]
51. Zhang, P.; Li, S.; Zhang, Z. General relationship between strength and hardness. *Mater. Sci. Eng. A* **2011**, *529*, 62. [CrossRef]
52. Caiazzo, F.; Alfieri, V.; Casalino, G. On the Relevance of Volumetric Energy Density in the Investigation of Inconel 718 Laser Powder Bed Fusion. *Materials* **2020**, *13*, 538. [CrossRef]
53. Rezende, M.C.; Araujo, L.S.; Gabriel, S.B.; dos Santos, D.S.; de Almeida, L.H. Hydrogen embrittlement in nickel-based superalloy 718: Relationship between $\gamma' + \gamma''$ precipitation and the fracture mode. *Int. J. Hydrogen Energy* **2015**, *40*, 17075–17083. [CrossRef]
54. Zhang, D.; Feng, Z.; Wang, C.; Wang, W.; Liu, Z.; Niu, W. Comparison of microstructures and mechanical properties of Inconel 718 alloy processed by selective laser melting and casting. *Mater. Sci. Eng. A* **2018**, *724*, 357–367. [CrossRef]
55. Zhang, Z.; Obasi, G.; Morana, R.; Preuss, M. In-situ observation of hydrogen induced crack initiation in a nickel-based superalloy. *Scr. Mater.* **2017**, *140*, 40–44. [CrossRef]
56. Sundararaman, M.; Mukhopadhyay, P.; Banerjee, S. Deformation behaviour of γ'' strengthened inconel 718. *Acta Metall.* **1988**, *36*, 847–864. [CrossRef]
57. Janssen, M.; Zuidema, J.; Wanhill, R. *Fracture Mechanics*; CRC Press: Boca Raton, FL, USA, 2004.
58. Lin, J.; Chen, F.; Liu, F.; Xu, D.; Gao, J.; Tang, X. Hydrogen permeation behavior and hydrogen-induced defects in 316L stainless steels manufactured by additive manufacturing. *Mater. Chem. Phys.* **2020**, *250*, 123038. [CrossRef]
59. Tarzimoghadam, Z.; Ponge, D.; Klöwer, J.; Raabe, D. Hydrogen-assisted failure in Ni-based superalloy 718 studied under in situ hydrogen charging: The role of localized deformation in crack propagation. *Acta Mater.* **2017**, *128*, 365–374. [CrossRef]

metals

Article

Microstructure and Mechanical Properties of Hastelloy X Fabricated Using Directed Energy Deposition

Yoon-Sun Lee and Ji-Hyun Sung *

Smart Manufacturing Technology R&D Group, Korea Institute of Industrial Technology, Daegu 42994, Republic of Korea
* Correspondence: jsung@kitech.re.kr; Tel.: +82-053-580-0146

Abstract: Laser-aided additive manufacturing is used for complex shapes and Ni-based superalloy parts. This study aimed to optimize the additive manufacturing process of Hastelloy X alloy to obtain its excellent mechanical properties without pores or cracks in the additively manufactured parts. The additively manufactured Hastelloy X was analyzed by comparing porosity, microstructure, and mechanical properties in as-built and post-heat treatment conditions. In addition, the pores existing inside the as-built specimen considerably decreased after the hot isostatic press (HIP) treatment. Furthermore, cell/columnar microstructures were observed owing to a fast cooling rate in the as-built condition. However, after heat treatment, dendrite structures disappeared, and recrystallized equiaxed grains were observed. The tensile test results showed that there was mechanical anisotropy along the vertical and horizontal directions, and as the microstructure changed to equiaxed grains after heat treatment, the mechanical anisotropy decreased, and the high-temperature properties improved.

Keywords: Hastelloy X; high-temperature properties; directed energy deposition

1. Introduction

Hastelloy X (HX) is a Ni-Cr-Fe-Mo-based heat-resistant alloy commonly used in gas turbine parts for aircraft, power plants, and petrochemical plants owing to its excellent corrosion resistance and high-temperature mechanical properties [1–3]. This alloy does not precipitate strengthening phases such as γ', and Cr and Mo act as solid solution-strengthening alloys, exhibiting solid solution-strengthening effects. Traditional casting methods are mainly used to produce HX parts. However, using these methods to manufacture complex HX parts causes long production cycles and increases costs. Recently, there have been an increasing number of cases in which additive manufacturing technology is applied to produce the necessary part shape [4,5].

Additive manufacturing (AM) is an advanced metal manufacturing process for the direct fabrication of near-net-shaped three-dimensional components from the solid model data using a high-power laser as a heat source. Until recently, many studies focusing on the development of metal powder materials for additive manufacturing and the optimization of the additive process have been conducted [6–9]. AM technologies for metals can be largely classified into directed energy deposition (DED) and powder bed fusion (PBF) processes. PBF is a process of depositing metal powders by laying them flat and irradiating a laser or electron beam on a designated area to melt the powders. DED is a process that melts and deposits materials using a high-power laser beam while directly supplying metal powders on the substrate. This method, which is similar to welding, can be applied to repair work because powders can be stacked on top of existing products. Moreover, the DED process can be used to produce alloys or deposit different materials using various metal powders because heterogeneous materials can be deposited.

Many studies have been conducted on HX AM using the PBF process [10–13], in contrast to the limited studies conducted on the DED process [14,15]. To apply the DED process, process optimization is required for the HX material. Among the various process parameters, laser beam power, laser scanning speed, and powder feeding rate have a significant impact on quality and process efficiency [16–19]. In addition, post-heat treatment may be required if necessary. These factors are directly related to the quality of the DED-manufactured materials. Therefore, the optimal process conditions must be chosen by controlling the DED process parameters and post-processing.

Regarding additively manufactured materials, residual stress, porosity, and anisotropy in mechanical properties are due to the high thermal gradients induced by rapid melting and cooling [20–22]. Therefore, ductility and mechanical properties can be significantly improved by applying stress-relief treatment to minimize residual stresses observed in as-built parts or by applying HIP (hot isostatic pressing) to remove pores or defects existing inside the part. Qiu et al. reported improved results after investigating the anisotropy of the ductility of Ti64 alloy manufactured with SLM (selective laser melting) in the as-built treatment and after HIP [23]. In addition, ductility was improved with a decrease in strength, but the anisotropy of ductility was maintained. Tomus et al. reported that pores and cracks can be generated during additive manufacturing, which can affect mechanical properties; however, this can be controlled by optimizing process parameters, including HIP [24].

In this study, the deposition of HX alloy was performed through laser-aided direct metal tooling (DMT), which is a DED process. An optimized process was developed that would result in a crack- and pore-free deposited layer, with excellent mechanical properties. The laser beam power and feeding rate were chosen as the most critical process parameters for this study. To investigate the anisotropy of the AM material, tension tests at room and high temperatures, as well as creep tests, were conducted in both horizontal and vertical directions. Additionally, the mechanical properties before and after the heat treatments were compared and analyzed.

2. Materials and Methods

2.1. Powder Properties

In this study, a commercial gas-atomized HX powder (AMC Powders Co. Ltd., Beijing, China) was used, with the alloy composition shown in Table 1. Particle size analysis was conducted using Matersizer 3000 (Malvern Panalytical, Malvern, UK), and the morphology of the powder was observed using scanning electron microscopy (FE-SEM, Hitachi, Tokyo, Japan). It had a spherical morphology, with a particle size range of approximately 57–145 μm (Figure 1). The mean D50 diameter was 92.1 μm. However, some powders had irregular shapes, with small satellites attached to them.

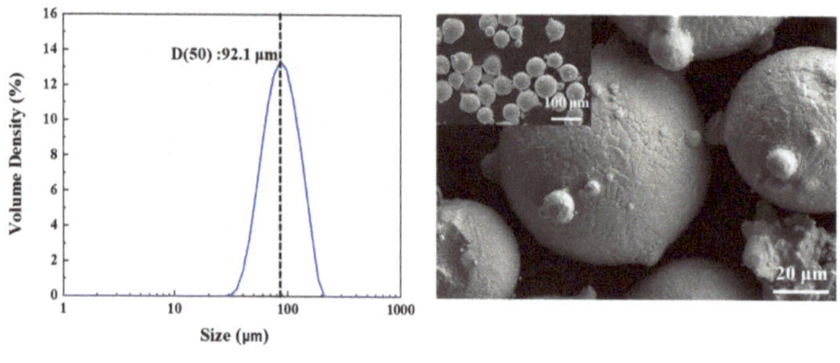

(a) (b)

Figure 1. Hastelloy X (HX) powder for direct energy deposition (DED): (**a**) particle size distribution HX powder; (**b**) SEM image of gas-atomized HX powder showing spherical shapes.

Table 1. Chemical composition of Hastelloy X.

Element (wt%)	Ni	Cr	Fe	Mo	W	Co	C	Si	Mn
Powder HX	Bal.	21.3	19.1	8.6	0.65	0.96	0.07	0.16	0.54

2.2. Direct Energy Deposition (DED) Process

The DED process involves deposition in real time during which a high-power laser beam is directed onto a metal surface, and metal powders are simultaneously fed with the generation of a molten pool. Thus, a metal layer with a dense structure is formed through a rapid melting and cooling process.

The equipment used in this study was a metal DED-type three-dimensional (3D) printer (MX-3, Insstek Inc., Daejeon, Republic of Korea). The laser of the used equipment was 2 kW Yb: YAG (IPG, Oxford, MA, USA), and the laser beam size was set to 1 mm. Powders were supplied through three powder hoppers, and argon was used for coaxial and powder carrier gases to prevent oxidation during the process. Figure 2 shows the DED system used for the deposition of the HX material.

The process experiment was performed with square specimens (10 mm × 10 mm × 5 mm) that were cut in the building direction and used in the microstructure analysis. The optimized process parameters are shown in Table 2.

Tensile and creep specimens were prepared to evaluate their mechanical properties. Furthermore, vertical and horizontal deposition directions were used to verify the differences in mechanical properties according to the building direction. Furthermore, heat treatment was performed to improve porosity and mechanical properties. Three different sample conditions (Table 3) were used: as-built treatment, heat treatment 1 (HT1), and heat treatment 2 (HT2). In the HT1 condition, hot isostatic pressing (HIP) was performed to improve porosity under the following conditions: 1150 °C, 4 h, and 100 MPa. In the HT2 condition, after HIP, solution heat treatment was additionally performed at 1177 °C for 30 min (solution heat treatment, SHT) in specimens to improve their properties.

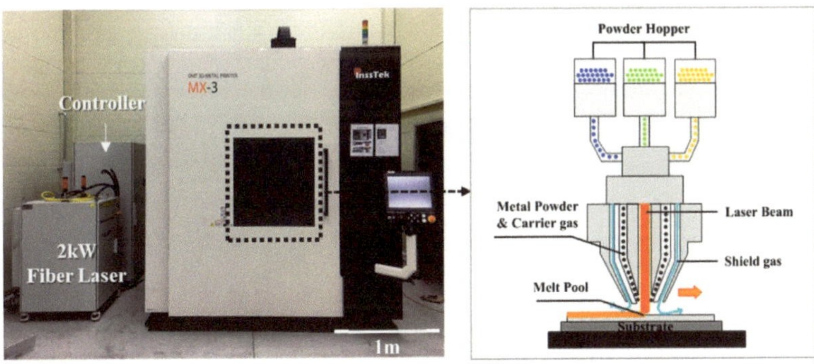

Figure 2. DED system used for the deposition of Hastelloy X alloy.

Table 2. DED process parameters for Hastelloy X alloy.

Power (W)	Scanning Speed (m/min)	Powder Feeding Rate (g/min)	Overlap Ratio (%)	Layer Thickness (mm)
450	0.85	3.5	60	0.25

Table 3. Summary of the three testing conditions.

Specimen	As-Built	HT1	HT2
Conditions	As-built	1150 °C, 4 h, 100 MPa (Furnace cooling)	1150 °C, 4 h, 100 MPa + 1177 °C, 30 min (N_2 gas cooling)

2.3. Microstructure Evaluation

For microstructure evaluation, the samples were ground and polished with 3 and 0.25 μm diamond suspensions. Thereafter, the prepared sample surface was immediately etched with a mixed solution consisting of 5 g $CuCl_2$ in 100 mL HCl and 100 mL CH_3CH_2OH (Kalling's No.2 reagent). Finally, the microstructure was observed using optical microscopy (OM, Axio Observer. D1m, Carl Zeiss, Oberkochen, Germany) and scanning electron microscopy (SEM, JXA-8530F, JEOL, Tokyo, Japan). X-ray diffraction (XRD) was performed using an X-ray diffractometer (SmartLab, RIGAKU, Tokyo, Japan), in which Cu Kα1 radiation was used, and the 2θ range was set from 20° to 100°. Electron backscattered diffraction (EBSD) analyses were performed using an FEI Quattro ESEM system (Thermo Fisher, Waltham, MA, USA) equipped with a Bruker QUANTAX EBSD detector at an acceleration voltage of 20 kV, a step size of 0.12 μm, and a working distance of 19.5 mm.

2.4. Evaluation of Mechanical Properties

As shown in Figure 3a, specimens for tensile and high-temperature creep tests were fabricated under the proposed optimal conditions of DED. The cube-shaped specimens were manufactured in vertical and horizontal directions to confirm anisotropy. The deposited specimens were processed into tensile and creep specimens using wire cutting. Tensile tests were conducted at room temperature (24 °C) with a strain rate of 8.0×10^{-4}/s for the as-built, HT1, and HT2 specimens fabricated in accordance with the ASTM E8 standard, as shown in Figure 3b (Z60, Zwick Roell, Ulm, Germany). Furthermore, tensile tests were conducted at a high temperature (816 °C) with a strain rate of 8.0×10^{-4}/s using the specimens fabricated in the same way. The high-temperature creep test specimens were fabricated in accordance with the ASTM E8 standard, as shown in Figure 3c, and tests were conducted under the conditions of 816 °C, 103 MPa, and an Ar environment according to the ASTM E139-11 test standard (RB 305, R&B, Daejeon, Republic of Korea).

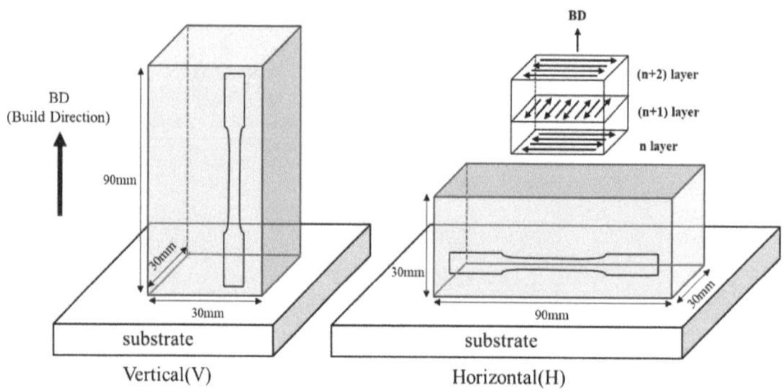

(a)

Figure 3. *Cont.*

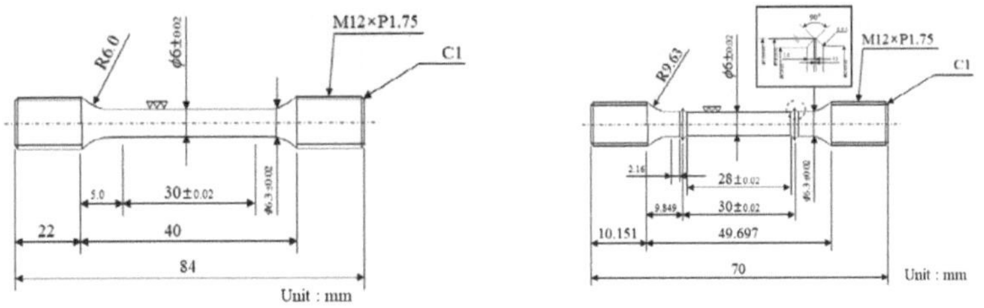

(b) (c)

Figure 3. Specimens for testing mechanical properties: (**a**) schematics of the HX samples according to the building direction and scanning strategy; (**b**) tensile test specimen specification; (**c**) creep test specimen specification.

3. Results and Discussion

3.1. Microstructure

As shown in Figure 4, the OM was used to detect the vertical cross-sectional porosity of the as-built and heat-treated specimens manufactured under optimal process conditions. Furthermore, the porosity was reduced after heat treatment, compared with the as-built specimen. The results of the cross-sectional porosity, maximum pore size, and pore number of the samples for each condition are summarized in Table 4. The cross-sectional porosity of the as-built specimens was 0.043%, whereas those of HT1 and HT2 specimens were 0.004% and 0.003%, respectively. The as-built specimens had spherical pores, which were assumed to have been generated from the gas trapped during the process. However, it was confirmed that the residual pores of the as-built specimens were significantly reduced after both heat treatments (HT1 and HT2), and the maximum pore size also decreased.

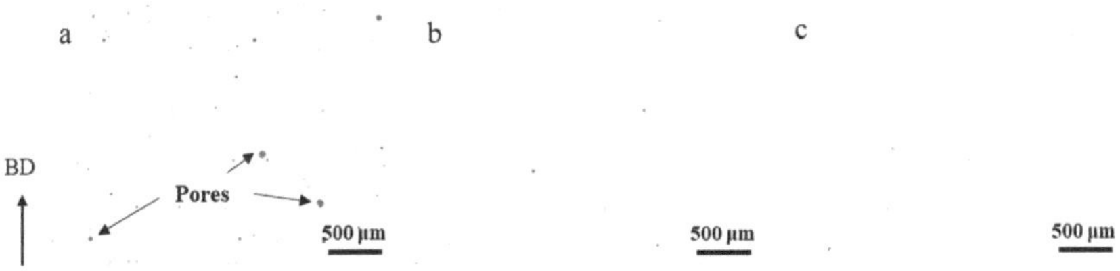

Figure 4. Cross-sectional porosity results of Hastelloy X in the building direction plane: (**a**) as-built, (**b**) HT1, and (**c**) HT2.

Table 4. Porosity measurement results of cross-sectional specimens.

Specimen	As-Built	HT1	HT2
Cross-sectional porosity (%)	0.043	0.004	0.003
Maximum pore size (μm)	93.626	27.563	11.666
Tracked Pore number	71	12	19

The cast HX alloy solidified relatively slowly, exhibiting a typical dendritic structure and coarse precipitates [25]. However, during the DED process, a complex microstructure was formed owing to the rapid melting and solidification process. The microstructures of the HX specimens fabricated through the DED process were observed, as shown in Figure 5. The microstructure of the as-built specimen exhibited a boundary of the molten pool (Figure 5a), which was generated due to the laser scan path during deposition. Additionally, the SEM images show the fine cell/columnar structure inside molten pool boundaries. The 3D structure of the cellular structure was similar to a honeycomb and was found to have different shapes depending on the direction of observation [26]. When observed transversely to the cell growth direction, an equiaxed cell structure was observed, and when observed in parallel to the cell growth direction, an elongated rod-type columnar structure was observed.

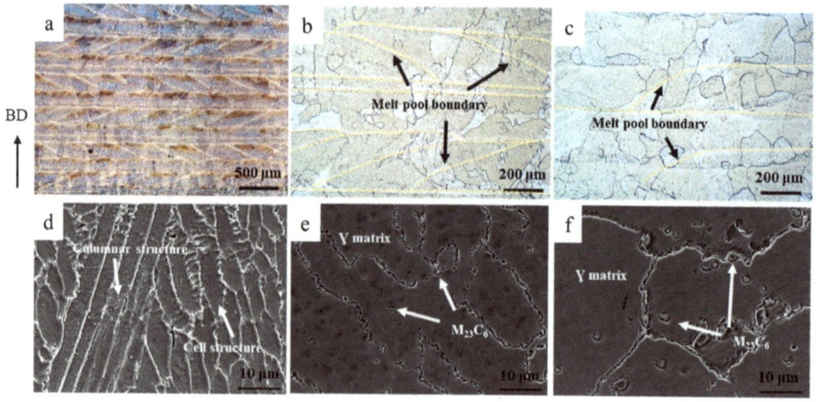

Figure 5. Optical micrographs and SEM images of HX specimens in the building direction plane: (**a**,**d**) as-built; (**b**,**e**) HT1; (**c**,**f**) HT2.

In the HT1 specimen, almost all the residual pores and dendrite structure disappeared, and recrystallized grains were formed through heat treatment, while film-shaped carbides were formed along the grain boundary, as shown in Figure 5b,e. In the case of the HT1 specimen subjected to isostatic pressure for a relatively long time, Cr-rich M23C6 carbide precipitated in the form of a film along the grain boundary. Conversely, in the HT2 specimen, with the addition of SHT, equiaxed grains were formed, and globular and square inter-/intragranular carbides were generated in the HT2 specimen, as shown in Figure 5c,f. Even after heat treatment, the molten pool boundary did not completely disappear. The existence of the molten pool boundary caused mechanical property anisotropy along the vertical/horizontal building direction.

SEM and EDS analyses confirmed the formation of Cr-rich $M_{23}C_6$ carbide at the grain boundary and inside the grains of the HT-1 and HT2 specimens, which underwent post-heat treatment. According to the literature, the secondary phases that can be generated in HX are carbides (M_6C and $M_{23}C_6$), σ, and μ phases [1,27]. However, the only precipitates inside the γ matrix of HX are carbides, unless aging is performed for an extended period [28]. Figure 6 shows the XRD patterns of the HX alloy's cross-sectional surface in the as-built and heat-treated samples. In the XRD patterns, the characteristic peaks of (1 1 1), (2 0 0), (2 2 0), (3 1 1), and (2 2 2) of γ phase nickel alloy with an FCC (face-centered cubic) crystal structure were detected. Low-intensity peaks associated with the carbide phase were also observed. These peaks can be attributed to $M_{23}C_6$ carbide dispersed in the γ matrix, as mentioned previously.

To confirm the grain size for each specimen, EBSD analysis was performed, and the results are shown in Figure 7. The grain size of the HT2 specimen was determined to be 217 μm based on the EBSD analysis. The grains of the as-built specimens were formed in a direction parallel to the building direction, which was the temperature-gradient direction. In the case of HT1 specimens subjected to isostatic pressure at a high temperature, recrystallization occurred, and it was confirmed that equiaxed grains were formed in HT2 specimens after SHT. This difference in grain shape caused variation in the mechanical properties depending on the building direction. The average grain size of each specimen was as follows: as-built (184 μm), HT1 (125 μm), and HT2 (217 μm). The grain size of the HT2 specimen was coarser than that of the HT1 specimen, at 125 μm. This indicates that, although the exposure to heat at 1177 °C for 30 min was sufficient for grain formation, it was not adequate to completely eliminate the molten pool boundary (MPB). Therefore, the heat treatment time was significantly short.

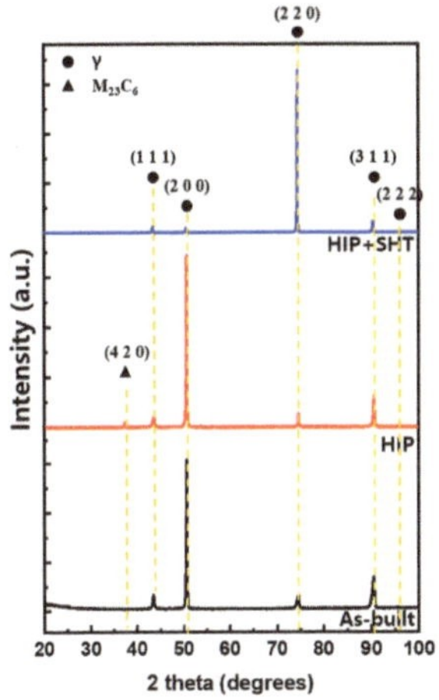

Figure 6. X-ray diffraction (XRD) patterns of as-built, HT1, and HT2 samples.

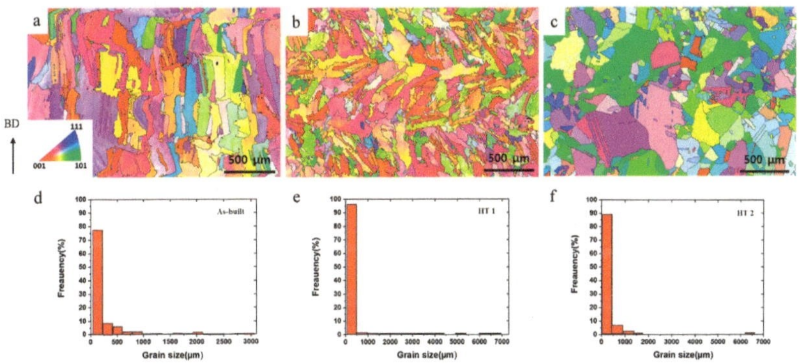

Figure 7. Electron backscatter diffraction (EBSD) results of HX specimens in the building direction plane (vertical): (**a**,**d**) as-built; (**b**,**e**) HT1; (**c**,**f**) HT2.

3.2. Tensile Properties

Figure 8 shows the results of tensile tests at room and high temperatures in as-built, HT1, and HT2 conditions with a vertical direction. In the room temperature test, the as-built specimen had a tensile stress of approximately 693 MPa and a uniform elongation of approximately 52.4%. After heat treatment, the tensile strength values of HT1 and HT2 were similar, but the yield strength significantly decreased (by approximately 228 MPa). However, elongation did not increase significantly.

Moreover, the high-temperature tensile test results revealed that the as-built specimen and the post-heat-treated specimen had similar tensile strength, but the yield strength decreased significantly (by approximately 90 MPa) after heat treatment (HT1 and HT2). However, although the elongation of the as-built specimen at a high temperature significantly decreased to 19%, elongation in HT1 and HT2 specimens significantly increased to 49% and 89%, respectively.

In contrast to the raw material, the HX specimen fabricated using AM had an MPB (Figure 5). Epitaxial growth caused a strong morphological texture in the building direction. Consequently, grains were formed and elongated in the vertical direction in the cross-sectional microstructure of the deposited specimen, as shown in Figure 7a. Moreover, cellular subgrains appeared in the grains. This cellular structure consists of high-density dislocation walls that act as dislocation barriers, significantly increasing the yield strength and hardness of the materials [11,24]. According to the literature, Mo-rich carbides appear in additively manufactured HX specimens [29]. However, in this study, Mo-rich carbides did not appear, and only Cr-rich carbides were observed after heat treatment.

As-built, HT1, and HT2 conditions affected the mechanical properties according to microstructural differences. As shown in Table 5, as-built specimens had a higher yield strength at room temperature. This is caused by high-density dislocation at subgrain boundaries. Conversely, HT1 and HT2 specimens revealed lower yield strength values than as-built specimens owing to the lower dislocation density resulting from the formation of recrystallized grains during heat treatment. The presence of Cr-rich $M_{23}C_6$ carbides along the grain boundaries and within the grains was also confirmed. After heat treatment, tensile strength and elongation values were similar to those of the raw material.

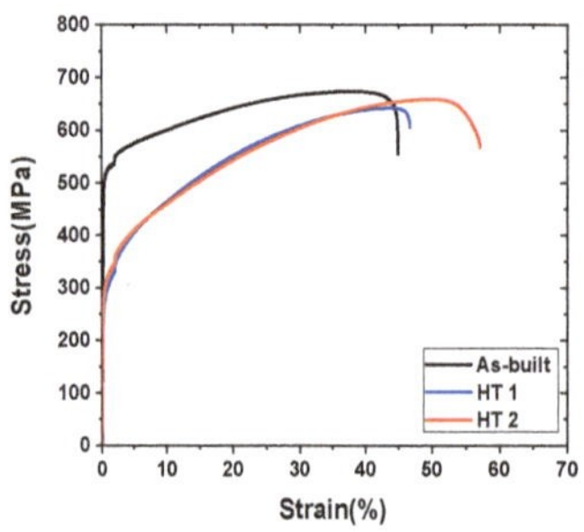

(a)

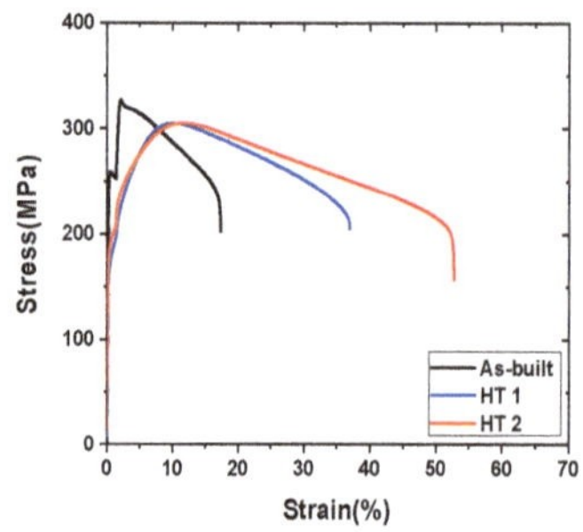

(b)

Figure 8. Stress–strain curves of the Hastelloy X alloy post-heat treatment in comparison to the as-built sample (vertical direction): (**a**) room temperature (24 °C); (**b**) high temperature (816 °C).

Table 5. Tensile test results of the as-built, HT1, and HT2 Hastelloy X specimens with different building directions: horizontal direction (H) and vertical direction (V).

Specimen		R.T. (24 °C) Y.S (MPa)	R.T. (24 °C) UTS (MPa)	R.T. (24 °C) Elongation (%)	H.T. (816 °C) Y.S (MPa)	H.T. (816 °C) UTS (MPa)	H.T. (816 °C) Elongation (%)
Raw		324 ± 1.4	739.5 ± 4.9	53.3 ± 2.8	210.5 ± 21.9	343 ± 1.4	67.2 ± 0.4
As -built	H	508 ± 0.0	566 ± 1.4	5.75 ± 1.1	244 ± 0.0	312.5 ± 2.1	3.5 ± 3.5
	V	524.5 ± 27.6	693 ± 26.9	52.4 ± 2.1	259.5 ± 3.5	328 ± 1.4	19.4 ± 2.3
HT1	H	361.5 ± 9.2	796.5 ± 4.9	30 ± 1.4	214.5 ± 0.7	348 ± 2.8	28.5 ± 3.5
	V	296.5 ± 46.0	672.5 ± 43.1	50.3 ± 0.7	166.5 ± 2.1	305.5 ± 0.7	49.2 ± 1.5
HT2	H	311 ± 9.9	758 ± 4.2	48.7 ± 7.2	197 ± 1.4	333.5 ± 3.5	96.6 ± 11.3
	V	294.5 ± 0.7	695.5 ± 50.2	57 ± 7.3	185 ± 0.0	309 ± 4.2	88.9 ± 37.3

The high-temperature tensile test results indicated that yield strength (YS) and ultimate tensile strength (UTS) decreased. According to the literature, this can be attributed to the decrease in the dislocation density as the sample is exposed to a higher temperature during testing [11]. In high-temperature tensile strength testing, grain boundary sliding (GBS) acts as a major factor causing fracture owing to the dislocation motion near grain boundaries [30]. Therefore, the larger the grain size, the higher the high-temperature strength. However, the as-built specimen showed the highest yield strength owing to the high dislocation density within the grain boundaries, even though the grain size was smaller than that of the HT2 specimen. Comparing the elongation of the different conditions, the HT2 specimen showed the highest elongation value. As shown in Figure 5f, discontinuous carbides were formed at the grain boundary, which suppressed dislocation movement along the grain boundary, resulting in high elongation.

Table 5 summarizes the tensile test results at room and high temperatures in vertical and horizontal directions. The anisotropy of mechanical properties according to the building direction appeared in all three conditions (as-built, HT1, and HT2). This anisotropy according to the building direction was caused by the molten pool boundaries generated in the microstructure of the as-built specimen (Figure 5). When fabricated in the vertical direction, the MPB was in the vertical direction at which the tensile force was applied, as shown in Figure 3a. The MPB of the vertical AM specimen also acted as a crack-propagation path because the cracks causing the fracturing of the specimen were formed in the vertical direction at which the tensile force was applied. In contrast, the horizontal AM specimen interfered with the propagation of cracks because the MPB was formed in the horizontal direction at which the tensile force was applied [31–34]. The strength of the horizontal specimen was higher than that of the vertical specimen, whereas the elongation showed the opposite result. In the case of the HT1 specimen, recrystallization commenced as the cell/columnar structures were removed, but the recrystallization process remained incomplete. In the case of the HT2 specimen, anisotropy according to the building direction was eliminated because, despite the remaining molten pool boundaries, equiaxed grains were uniformly generated throughout the specimen as recrystallization proceeded. When the results were compared based on the building direction, the HT2 specimen showed the lowest anisotropy effect, and the tensile properties of the HT2 specimens were the most similar to those of the raw material.

3.3. Analysis of High-Temperature (816 °C) Creep Rupture Behavior

Figure 9 shows the high-temperature creep test results according to the building direction for as-built, HT1, and HT2 conditions. The result of the high-temperature creep test showed different creep properties according to the building direction and heat treatment. With respect to the creep property, the rupture time of the vertical direction in the as-built condition was 33 h, approximately 83% larger than that of the horizontal direction (18 h).

Regarding the high-temperature creep property, GBS was the primary cause of the rupture. For the as-built specimens, grains were formed parallel to the building direction as a result of which heat was released. The specimen fabricated in the vertical direction exhibited a higher creep life than that built in the horizontal direction because the area of GBS was minimized by the vertically elongated grains. The columnar grain morphology caused the anisotropy of the creep properties in both specimens. In the horizontal specimen, the grain boundary was aligned normally to the loading axis, reducing the creep life [35]. In addition, as no carbide was present at the grain boundary, the voids adhered to each other when deformed at high temperatures, resulting in excessive GBS. Consequently, the as-built creep life was low.

After heat treatment, the dendritic structure began to disappear, and equiaxed grains and carbides appeared. This study confirmed that the specimen of HT2 with a large grain size had the best creep properties after heat treatment (V: 91 h, H: 114 h). The grain formation due to heat treatment minimized the area of the grain boundary. Moreover, as carbide was unevenly distributed at the grain boundary, the grain boundary was strengthened, and the creep life improved [36–38]. However, as mentioned in Section 3.1, the homogenization of the microstructure was not sufficiently advanced; thus, the creep property still had anisotropy.

Figure 10 shows the fracture surface observed through SEM after the creep test. In the as-built specimen, a fracture along the dendritic structure was observed, as shown in Figure 10a. The fracture occurred in the "track–track" MPB region because the interdendritic fracture surface sloped into a structure resembling a "valley." In contrast, dimples, a typical ductile fracture mode, were observed in the heat-treated specimens.

From the high-temperature creep test results, the HT2 specimen exhibited the best creep properties among the as-built, HT1, and HT2 conditions.

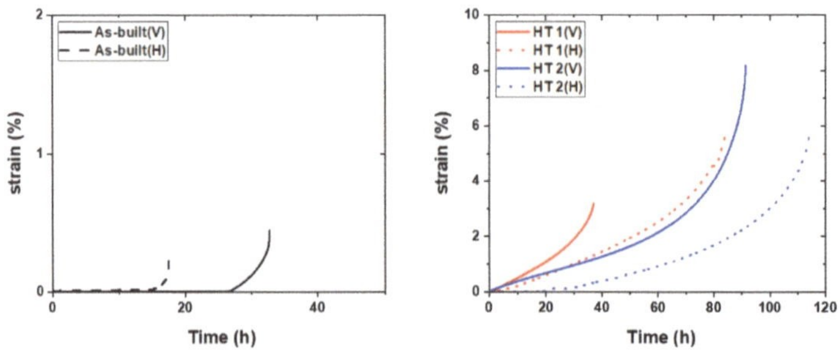

(**a**) (**b**)

Figure 9. Creep test results according to the building direction: (**a**) as-built; (**b**) heat treatment.

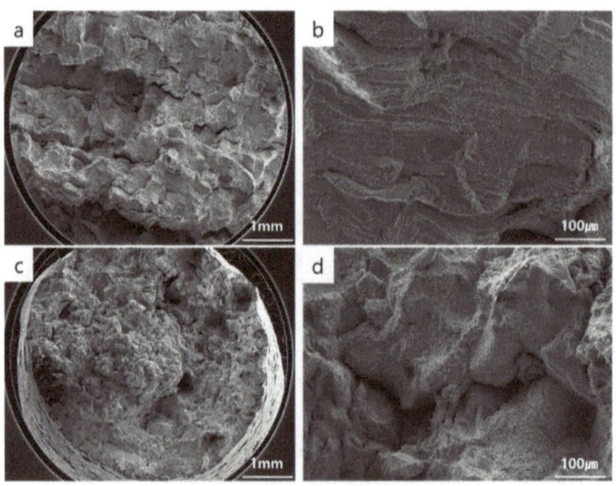

Figure 10. Creep fracture surface morphology of Hastelloy X specimens at various magnifications: (**a**,**b**) as-built; (**c**,**d**) HT1.

4. Conclusions

In this study, the optimized processing conditions for Hastelloy X alloy using the DED process were proposed, and changes in the microstructure and mechanical properties before and after heat treatment were compared. The results of this study are as follows:

(1) In the as-built specimens, spherical pores were generated owing to the trapped gas during the DED process. However, the residual pores significantly decreased after heat treatment. As-built specimens had a fine cell/columnar structure. After heat treatment, recrystallization occurred as the temperature increased, equiaxed grains were formed, and Cr-rich $M_{23}C_6$ carbides were generated in the γ matrix.

(2) At room temperature, the as-built specimens exhibited a high tensile strength because of the high-density dislocation at subgrain boundaries. Conversely, HT1 and HT2 specimens showed a lower yield strength than as-built specimens because of a lower dislocation density as recrystallized grains were formed because of heat treatment. The high-temperature tensile tests indicated that YS and UTS decreased.

(3) The anisotropy of mechanical properties according to the building direction appeared in all three conditions (as-built, HT1, and HT2). The strength of the horizontal

specimen was higher than that of the vertical specimen, and elongation showed the opposite result. When the results were compared according to the building direction, the HT2 specimen showed the lowest anisotropy effect, and the tensile properties of the HT2 specimen were the most similar to those of the raw material.

(4) GBS is the main cause of failure in high-temperature creep properties. For the as-built specimen, grains were formed parallel to the building direction as a result of which heat was released. The specimen fabricated in the vertical direction exhibited a higher creep life than that built in the horizontal direction because the area of GBS was minimized by the vertically elongated grains. After heat treatment, the dendritic structure began to disappear, and equiaxed grains and carbides appeared. This study confirmed that the HT2 specimen with a large grain size had the best creep properties after heat treatment.

(5) The MPB did not completely disappear even after HT2, because a temperature of 1177 °C for 30 min for SHT was not sufficient. Furthermore, anisotropy did not completely disappear in the high-temperature creep results.

(6) These research results highlight the superior high-temperature properties of Hastelloy X material that can be obtained using the DED process. However, further research is needed to completely eliminate anisotropy.

Author Contributions: Conceptualization, Y.-S.L. and J.-H.S.; methodology Y.-S.L. and J.-H.S.; validation, J.-H.S.; investigation, Y.-S.L. and J.-H.S.; data curation, Y.-S.L.; writing—original draft preparation, Y.-S.L.; writing—review and editing, Y.-S.L. and J.-H.S.; visualization, Y.-S.L.; supervision, J.-H.S.; project administration, J.-H.S. All authors have read and agreed to the published version of the manuscript.

Funding: This work was supported by the Korea Institute of Energy Technology Evaluation and Planning (KETEP) and the Ministry of Trade, Industry, and Energy (MOTIE) of the Republic of Korea (No. 20206700000010).

Institutional Review Board Statement: Not applicable.

Informed Consent Statement: Not applicable.

Data Availability Statement: Not applicable.

Conflicts of Interest: The authors declare no conflict of interest.

References

1. Zhao, J.C.; Larsen, M.; Ravikumar, V. Phase precipitation and time–temperature-transformation diagram of Hastelloy X. *Mater. Sci. Eng. A* **2000**, *293*, 112–119. [CrossRef]
2. Haynes-International. *Hastelloy X Alloy*; Haynes-International: Kokomo, IN, USA, 1997; Volume 16.
3. Hong, H.U.; Kim, I.S.; Choi, B.G.; Jeong, H.W.; Jo, C.Y. Effects of temperature and strain range on fatigue cracking behavior in Hastelloy X. *Mater. Lett.* **2008**, *62*, 4351–4353. [CrossRef]
4. Harrison, N.J.; Todd, I.; Mumtaz, K. Reduction of micro-cracking in nickel superalloys processed by selective laser melting: A fundamental alloy design approach. *Acta Mater.* **2015**, *94*, 59–68. [CrossRef]
5. Wilson, J.M.; Piya, C.; Shin, Y.C.; Zhao, F.; Ramani, K. Remanufacturing of turbine blades by laser direct deposition with its energy and environmental impact analysis. *J. Clean.* **2014**, *80*, 170–178. [CrossRef]
6. Trojan, K.; Ocelík, V.; Capek, J.; Cech, J.; Canelo-Yubero, D.; Ganev, N.; Kolarík, K.; De Hosson, J.T.M. Microstructure and Mechanical Properties of Laser Additive Manufactured H13 Tool Steel. *Metals* **2022**, *12*, 243. [CrossRef]
7. Cao, L.; Zhang, L.; Bi, G. Feasibility Study on Deposition of Tribaloy T800 on Cobalt-Based L605 Using Micro-Laser-Aided Additive Manufacturing. *Metals* **2022**, *12*, 586. [CrossRef]
8. Vu, H.M.; Meiniger, S.; Ringel, B.; Hoche, H.C.; Oechsner, M.; Weigold, M.; Schmitt, M.; Schlick, G. Investigation of Material Properties of Wall Structure from Stainless Steel 316L Manufactured by Laser Powder Bed Fusion. *Metals* **2022**, *12*, 285. [CrossRef]
9. Aversa, A.; Marchese, G.; Bassini, E. Directed Energy Deposition of AISI 316L Stainless Steel Powder: Effect of Process Parameters. *Metals* **2021**, *11*, 932. [CrossRef]
10. Wang, F. Mechanical property study on rapid additive layer manufacturing Hastelloy X alloy by selective laser melting technology. *Int. J. Adv. Manuf. Technol.* **2012**, *58*, 545–551. [CrossRef]

11. Montero-Sisroaga, M.L.; Liu, Z.; Bautmans, L.; Nardone, S.; Ji, G.; Kruth, J.-P.; Van Humbeeck, J.; Vanmeensel, K. Effect of temperature on the microstructure and tensile properties of micro-crack free hastelloy X produced by selective laser melting. *Addit. Manuf.* **2020**, *31*, 100995. [CrossRef]
12. Marchese, G.; Bassini, E.; Aversa, A.; Lombardi, M.; Ugues, D.; Fino, P.; Biamino, S. Microstructural Evolution of Post-Processed Hastelloy X Alloy Fabricated by Laser Powder Bed Fusion. *Materials* **2019**, *12*, 486. [CrossRef] [PubMed]
13. Li, Y.; Qi, H.; Hou, H.; Lei, L. Effects of Hot Isostatic Pressing on Microstructure and Mechanical Properties of Hastelloy X Samples Produced by Selective Laser Melting. *Adv. Eng. Res.* **2017**, *102*, 31–40. [CrossRef]
14. Zhang, S.; Liu, J.; Lin, X.; Huang, Y.; Zhang, Y.; Guo, P.; Li, J.; Huang, W. Microstructure and anodic electrochemical behavior of additive manufactured Hastelloy X alloy via directed energy deposition. *Addit. Manuf.* **2021**, *39*, 101824. [CrossRef]
15. Xu, L.; Gao, Y.; Zhao, L.; Han, Y.; Jing, H. Ultrasonic micro-forging post-treatment assisted laser directed energy deposition approach to manufacture high-strength Hastelloy X superalloy. *J. Mater. Process. Technol.* **2022**, *299*, 117324. [CrossRef]
16. Mahamood, R.M.; Akinlabi, E.T.; Shukla, M.; Pityana, S. Laser Metal Deposition of Ti6Al4V: A Study on the Effect of Laser Power on Microstructure and Microhardness. In Proceedings of the International MultiConference of Engineering and Computer Scientists (ICMCS 2013), Hong Kong, China, 13–15 March 2013.
17. Kim, J.S.; Kang, B.J.; Lee, S.W. An experimental study microstructure characteristics and mechanical properties of stainless-steel 316L parts using directed energy deposition (DED) process. *J. Mech. Sci. Technol.* **2019**, *33*, 5731–5737. [CrossRef]
18. Mathoho, I.; Akinlabi, E.T.; Arthur, N.; Tlotleng, M. Impact of DED process parameters on the metallurgical characteristics of 17-4PH SS deposited using DED. *CIRP J. Manuf. Sci. Technol.* **2020**, *31*, 450–458. [CrossRef]
19. Jinoop, A.N.; Paul, C.P.; Bindra, K.S. Laser assisted direct energy deposition of Hastelloy X. *Opt. Laser Technol.* **2019**, *109*, 14–19. [CrossRef]
20. Woo, I.Y.; Lyu, M.Y. Variations in the Impact strength of material extrusion-type 3D printed specimens depending on tool path and building direction. *Polym. Korea* **2020**, *44*, 471–478. [CrossRef]
21. Carroll, B.E.; Palmer, T.A.; Beese, A.M. Anisotropic tensile behavior of Ti-6Al-4V components fabricated with directed energy deposition additive manufacturing. *Acta Mater.* **2015**, *87*, 309–320. [CrossRef]
22. Hrabe, N.; Gnäupel-Herold, T.; Quinn, T. Fatigue properties of a titanium alloy (Ti–6Al–4V) fabricated via electron beam melting (EBM): Effects of internal defects and residual stress. *Int. J. Fatigue* **2017**, *94*, 202–210. [CrossRef]
23. Qiu, C.; Adkins, N.J.E.; Attallah, M.M. Microstructure and tensile properties of selectively laser-melted and of HIPed laser-melted Ti–6Al–4V. *Mater. Sci. Eng. A* **2013**, *578*, 230–239. [CrossRef]
24. Tomus, D.; Tian, Y.; Rometsch, P.A.; Heilmaier, M.; Wu, X. Influence of post heat treatment on anisotropy of mechanical behavior and microstructure of Hastelloy-X parts produced by selective laser melting. *Mater. Sci. Eng. A* **2016**, *667*, 42–53. [CrossRef]
25. Choi, B.G.; Kim, I.S.; Do, J.H.; Jung, J.E.; Jung, I.Y.; Hong, H.U.; Jo, C.Y. The Effect of Thermal Exposure on the Microstructural Evolution and Tensile Properties in Cast Hastelloy X. *J. Korea Foundry Soc.* **2017**, *37*, 139–147. [CrossRef]
26. Kong, D.; Dong, C.; Wei, S.; Ni, X.; Zhang, L.; Li, R.; Wang, L.; Man, C.; Li, X. About metastable cellular structure in additively manufactured austenitic stainless steels. *Addit. Manuf.* **2021**, *28*, 101804. [CrossRef]
27. Jiang, L.; Hu, R.; Kou, H.; Li, J.; Bai, G.; Fu, H. The effect of M23C6 carbides on the formation of grain boundary serration in a wrought Ni-based superalloy. *Mater. Sci. Eng. A* **2012**, *536*, 37–44. [CrossRef]
28. Tawancy, H.M. Long-term ageing characteristics of Hastelloy alloy X. *J. Mater.* **1983**, *18*, 2976–2986. [CrossRef]
29. Marchese, G.; Basile, G.; Bassini, E.; Aversa, A.; Lombardi, M.; Ugues, D.; Fino, P.; Biamino, S. Study of the microstructure and cracking mechanisms of hastelloy X Produced by laser powder bed fusion. *Materials* **2018**, *11*, 106. [CrossRef]
30. Jang, J.E.; Kim, Y.S.; Sung, J.H.; Kim, Y.J.; Park, S.H.; Kim, D.H. Microstructural control strategy based on optimizing laser powder bed fusion for different Hastelloy X powder size. *Materials* **2022**, *15*, 6191. [CrossRef]
31. Syed, A.K.; Ahmad, B.; Guo, H.; Machry, T.; Eatock, D.; Meyer, J.; Fitzpatrick, M.E.; Zhang, X. An experimental study of residual stress and direction-dependence of fatigue crack growth behaviour in as-built and stress-relieved selective-laser-melted Ti6Al4V. *Mater. Sci. Eng. A* **2019**, *755*, 246–257. [CrossRef]
32. Wauthle, R.; Vrancken, B.; Beynaerts, B.; Jorissen, K.; Schrooten, J.; Kruth, J.-P.; Van Humbeeck, J. Effects of build orientation and heat treatment on the microstructure and mechanical properties of selective laser melted Ti_6Al_4V lattice structures. *Addit. Manuf.* **2015**, *5*, 77–84. [CrossRef]
33. Tao, P.; Li, H.-X.; Huang, B.-Y.; Hu, Q.-D.; Gong, S.-L.; Xu, Q.-Y. Tensile behavior of Ti-6Al-4V alloy fabricated by selective laser melting: Effects of microstructures and as-built surface quality. *China Foundry* **2018**, *15*, 243–252. [CrossRef]
34. Mahmud, A.; Huynh, T.; Zhou, L.; Hyer, H.; Mehta, A.; Imholte, D.D.; Woolstenhulme, N.E.; Wachs, D.M.; Sohn, Y. Mechanical Behavior Assessment of Ti-6Al-4V ELI Alloy Produced by Laser Powder Bed Fusion. *Metals* **2021**, *11*, 1671. [CrossRef]
35. Banoth, S.; Palleda, T.N.; Saito, T.; Murakami, H.; Kakehi, K. Effect of yttrium and silicon contents in Hastelloy-X built by selective laser melting process. *J. Alloys Compd.* **2022**, *896*, 163050. [CrossRef]
36. Thébaud, L.; Villechaise, P.; Crozet, C.; Devaux, A.; Béchet, D.; Franchet, J.M.; Rouffie, A.L.; Mills, M.; Cormier, J. Is there an optimal grain size for creep resistance in Ni-based disk superalloys? *Mater. Sci. Eng. A* **2018**, *716*, 274–283. [CrossRef]

37. Bang, J.H.; Kang, Y.J.; Kim, N.K.; Seo, S.M.; Lee, S.H.; Song, S.W.; Kang, N.H. Effect of aging heat treatment conditions on the mechanical properties and microstructure of base and weld metal of alloy 282 superalloy. *Korean J. Met. Mater.* **2020**, *8*, 540–549. [CrossRef]
38. Wang, L.; Wang, D.; Liu, T.; Li, X.W.; Jiang, W.G.; Zhang, G.; Lou, L.H. Effect of minor carbon additions on the high-temperature creep behavior of a single-crystal nickel-based superalloy. *Mater. Char.* **2015**, *104*, 81–85. [CrossRef]

metals

MDPI

Article

Mechanical Properties and Fracture Behavior of Laser Powder-Bed-Fused GH3536 Superalloy

Haohan Ni [1,†], Qi Zeng [2,3,†], Kai Zhang [4], Yingbin Chen [1] and Jiangwei Wang [1,5,*]

1 Center of Electron Microscopy, State Key Laboratory of Silicon Materials, School of Materials Science and Engineering, Zhejiang University, Hangzhou 310027, China; haohan_ni@zju.edu.cn (H.N.); yingbin_chen@zju.edu.cn (Y.C.)
2 School of Aeronautics and Astronautics, Zhejiang University, Hangzhou 310027, China; 12024082@zju.edu.cn
3 AECC Hunan Aviation Plant Research Institute, Zhuzhou 412002, China
4 School of Materials and Chemistry, University of Shanghai for Science and Technology, Shanghai 200093, China; kai.zhang@monash.edu
5 Wenzhou Key Laboratory of Novel Optoelectronic and Nano Materials, Institute of Wenzhou, Zhejiang University, Wenzhou 325006, China
* Correspondence: jiangwei_wang@zju.edu.cn
† These authors contributed equally to this work.

Abstract: Heat treatment (HT) is an important approach to tune the structure and mechanical properties of as-printed or hot-isostatic-pressed (HIPed) additive manufacturing materials. Due to the carbide precipitates extensively existing after HT with air cooling, this paper studies the microstructure and mechanical behavior of laser powder-bed-fused (L-PBFed) GH3536 superalloy with laminar carbide precipitates at grain boundaries. By comparing with air-cooling samples and water-quenched samples, the results revealed that air cooling often introduced precipitates at grain boundaries, which impede the plastic deformation and are prone to lead to severe transgranular cracks on the fracture surface, contributing to a higher strain-hardening rate but lower ductility of HTed sample. Water quench can largely eliminate the grain-boundary precipitates, contributing to an optimized ductility even with smaller grain size. This work provides more details on the precipitate-deformation relation after HT.

Keywords: additive manufacturing; heat treatment; nickel-based superalloy; precipitates

1. Introduction

Nickel-based superalloys are the most widely used metallic materials as hot parts of gas turbine [1,2]. The GH3536 superalloy (Hastalloy X) is a typical solid-solution-strengthened nickel-based superalloy with high oxidation and corrosion resistances as well as superior mechanical properties at high temperature [3]. However, conventional manufacturing methods such as forging and casting, which were used to build hot parts, are cumbersome in steps and time-/material-consuming [4]. Meanwhile, melt defects, including carbides and segregations [5,6], hinder the improvement of the high-temperature performance of as-built GH3536 superalloys. Since 3D-printing technology has shown great potential in replacing or optimizing conventionally industrial manufacturing technologies, applying laser-additive manufacturing in building superalloy has been also wildly investigated [7].

Laser powder-bed fusion (L-PBF) is one of the laser-based additive manufacturing methods with the capability of higher shape precision, lower material waste, and fewer processing steps than conventional manufacturing technologies [8,9]. The mechanical properties of as-printed samples are often reported as higher-strength but lower-ductility compared to the casting or wrought parts [10,11]. Although a good combination of strength and ductility can be achieved in some metals and alloys, the spatial heating and cooling cycle during L-PBF often induce nonequilibrium and anisotropic structures [12,13], as well as high-density dislocations, which greatly impair the mechanical properties of as-printed materials. For example, the L-PBF-induced molten pool boundaries, crystallographic texture, micropores, and cracks can lead to low ductility and mechanical anisotropy of L-PBFed materials [13–16]. To eliminate these drawbacks, numerous efforts have been devoted to fabricating materials with excellent properties. Generally, by optimizing the parameters of the L-PBF process, including laser power [17], scan speed [18,19], and scan strategy [20,21], an enhanced relative density and weakened texture can be obtained. However, some metallurgical defects such as solidification cracking [22], liquation cracking [23], and pores [17] are difficult to eliminate via the optimization of printing parameters. Therefore, post-treatments, including heat treatment (HT) and hot isostatic pressing (HIP) are often introduced to reduce metallurgical defects of additive manufacturing for an optimized mechanical property.

After L-PBF fabrications, an anisotropic structure with high density of dislocations and metallurgical defects is formed because of the effects of thermal gradients, heat cycling, and solidification conditions. Fine structures with numerous dislocations lead to high tensile strength, but the anisotropy and metallurgical defects often cause limited ductility of L-PBFed Hastalloy X [24–26]. Since hot parts are sensitive to pores and cracks due to stress concentration especially under fatigue loading [27,28], HIP is often applied to homogenize the microstructure and reduce the metallurgical defects of L-PBFed metallic parts. However, carbide precipitates in grain interiors and grain boundaries after HIP have been extensively reported in previous studies, and these precipitated particles at grain boundaries often result in lower mechanical properties and thermal stability of L-PBFed Hastalloy X [29,30]. Thus, HIP is commonly combined with HT to simultaneously eliminate pores and cracks and dissolve the precipitates introduced by HIP. Noting that HT could also introduce precipitates [31] and that the porosity of the HIPed sample is prone to regrow after HT [32,33], it is necessary to reconsider the role of HT and to optimize HT parameters to achieve excellent mechanical properties.

Generally, during HT at solid-solution temperature, the residual stress, strong texture, submicrometric cell structures, and part of metallurgical defects can be greatly eliminated, contributing to an enhanced elongation and relative density of L-PBFed Hastalloy X [29,34]. However, the grain size and the density of precipitates are strongly correlated with HT parameters. Conventionally, the HT of the solid-solution alloy is mainly affected by homogenizing time and cooling conditions. Previous studies [29,30] tended to conduct a sufficient homogenizing time (e.g., 2 h) with air cooling to make full recrystallization and solid solution of Hastalloy X when merely applying HT. However, in the case of HIP combining with HT, a shorter homogenizing time with water quench was applied in the HIPed sample [35]. Hence, the mechanical properties and fracture behavior of Hastalloy X after different HT processes need further investigations in order to establish a full structure–property relation.

In this work, we investigate the mechanical behavior of L-PBFed and HTed GH3536 superalloy. Mechanical properties were measured at room temperature, and the influences of carbide precipitates at grain boundaries were discussed by comparing with samples under different HT conditions. The deformation mechanisms of as-printed and HTed samples were investigated. This paper reports the HT optimization for improving ductility of HTed metal parts and provides more details on the precipitate-deformation relation after HT.

2. Materials and Methods

GH3536 powders (sizes ranged from 10–53 μm, ordered from Avimetal Powder Metallurgy Technology Co., Ltd., Beijing, China) were used in this study. The nominal composition of as-received powders and the compositions of as-printed sample confirmed by energy-dispersive spectroscopy (EDS) quantification at beam energy of 20 kV are shown in Table 1. Because the elements with atomic number of less than 11 are not reliable in EDS, carbon and boron were not included. At least 10 points of EDS scanning were acquired and the molten-pool boundaries were avoided. L-PBF fabrication was carried out in EOS M290. The parameter optimization was carried out prior to this study by using Doelhert parameter-optimization approach (with the details found in [36]), with the initial parameter ranges as follows: laser power 195–360 W, laser scan speed 930–1200 mm/s, and hatch distance 0.08–0.13 mm. The layer thickness was fixed as 40 μm. Four rounds of parameter optimization with 60 samples were applied. During the L-PBF process, the printing parameters were set as laser power 245 W, scan speed 1198.6 mm/s, hatch distance 0.08 mm, and layer thickness 0.04 mm. A substrate preheating of 80 °C was used for the L-PBF process to reduce the residual stress generated during the fabrication process. After fabrication, the relative density of the as-fabricated sample was measured to be 99.8% based on the image-characterization method described in the previous study [37].

Table 1. Nominal and measured (EDS quantification) compositions of L-PBFed GH3536 superalloy.

Element	Ni	Cr	Fe	Mo	Co	Mn	W	Si	C	B
Nominal wt.%	Bal.	22.0	18.0	9.0	1.5	Max 1.0	0.6	Max 1.0	0.1	Max 0.008
Measured wt.%	Bal.	21.93 ± 0.17	18.83 ± 0.16	9.95 ± 0.45	1.57 ± 0.04	0.13 ± 0.12	0.84 ± 0.14	0.06 ± 0.04	—	—

Subsequently, we tried the different times of HT with air cooling and water quenching, including 0.5/1.0/1.5/2.0 h. We found that two hours' HT followed by air cooling and one hour followed by water quenching enabled the samples' best mechanical properties and similar grain size. Thus, these two samples were used in this paper: (1) homogenized and recrystallized at 1177 °C for 2 h followed by air cooling (1177-AC) in an indoor environment; (2) annealed at 1177 °C for 1 h followed by water quenching (1177-WQ). All the samples were heated with the furnace at a rate of 10 °C/min and without protection from atmosphere. After heat treatments, samples for microstructural characterizations and uniaxial tensile testing were cut perpendicular to the building direction of the L-PBF samples. The average grain size was determined by a linear intercept method according to the ASTM E112 standard (excluding the annealing twin boundaries (TBs)), and three images were counted in every HTed sample. For tensile testing, dog-bone samples with the gauge sections of 2.0 × 2.0 × 9.5 mm^3 were machined perpendicular to building direction by electrodischarge, followed by mechanical gridding with SiC paper. Room-temperature tensile tests were conducted in a screw-driven SUST-CMT5000GL mechanical testing machine at a strain rate of 10^{-3} s^{-1}.

For microstructural characterizations, samples for optical microscope observation were firstly cut horizontal to building direction and carefully ground by 200, 400, 800, 1200, and 2000 grit SiC paper in turn. Then two-step mechanical polishing by diamond-particle suspension with a particle size of 5 μm and 1 μm was conducted to make the surface of the sample as bright as a mirror. Samples for scanning electron microscope observation were further electro-etched in an Oxalic-acid saturated solution with a voltage potential of 6 V for 15–20 s at room temperature [24]. Hitachi SU-70 field-emission scanning electron microscope (SEM) equipped (Hitachi, Tokyo, Japan) with an Oxford energy-dispersive X-ray spectroscopy (EDS) detector (Oxford Instruments, Oxfordshire, United Kingdom) was employed to analyze the microstructure and element distribution. Transmission electron microscopy (TEM) samples were punched as discs with a diameter of 3 mm, and then twinjet-electropolished in a methanol solution containing 5 vol.% perchloric acid at −20 °C. Scanning TEM (STEM) images were carried out using a spherical aberration-corrected FEI Titan G2 80–200 ChemiSTEM (Thermo Fisher Scientific, Hillsboro, OR, United States) operating at 200 kV equipped with a high-angle annular dark-field (HAADF) detector and a bright-field (BF) detector. X-ray diffraction (XRD) was performed on a copper target (λ = 1.5406 Å) PANalytical X'Pert PRO system (PANalytical B.V., Almelo, The Netherlands) with generator voltage 40 kV and tube current 40 mA to determine the crystal structure of GH3536 superalloy. The scan-step size was 0.02 degrees.

3. Results

3.1. Microstructures

With the process of parameter optimization using the Doehlert experimental design approach, a dense GH3536 superalloy was fabricated by L-PBF. Figure 1 shows the high densification of the as-fabricated part horizontal to the building direction with no visible crack or print defect. Figure 2 shows the representative microstructures of the sample perpendicular to the building direction. As shown in Figure 2a, the as-fabricated samples contain numerous overlapped melting pools. These melting pools consist of multiple grains defined by the orientation of dendrites (Figure 2b). Bright-field STEM characterization shows that each grain in the melting pool contains many polygonal dislocation cells, with an average diameter of ~0.6 μm (Figure 2c). Figure 2d shows that the boundaries of these dislocation cells are composed of numerous tangled dislocations, with some residual precipitated particles distributed randomly in the interior and at the boundary of the dislocation cell. EDS mapping of this region further demonstrates that the precipitated particles mainly consist of Cr and Mo, while the segregated elements at cell boundaries contain Fe, Cr, Mo, and Mn (Figure 2e). According to previous studies [38,39], dislocation structures and precipitates are formed due to the rapid heating and cooling during printing, where the density of dislocations increases with thermal distortion and is affected by geometric constraints.

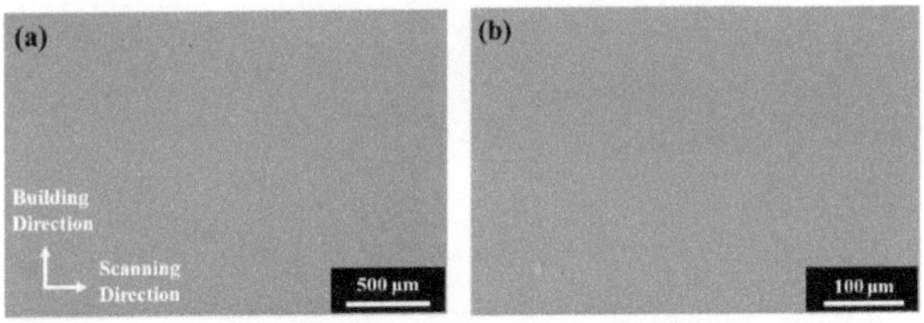

Figure 1. Optical micrographs of the as-fabricated GH3536 superalloy horizontal to building direction. (**a**) in low magnification, (**b**) in middle magnification.

Subsequently, the as-fabricated GH3536 superalloy was heat-treated under different conditions. Figure 3a,b show the microstructures of HTed samples, including 1177-AC (Figure 3a) and 1177-WQ (Figure 3b). After HT, the large melting-pool boundaries and dendrites were largely dissolved and the whole sample developed to a fully recrystallized state. The average grain sizes of 1177-AC and 1177-WQ samples were measured to be ~84 μm and ~65 μm, respectively. In the 1177-AC sample, a mass of precipitates formed inside the grains and at the grain boundaries (Figure 3a). This is because the air-cooling process can provide adequate time for element diffusion and thus induce segregation/precipitation. Some of the precipitates may also come from the residual ones that were not fully dissolved during HT. In the grain interior, the diameters of the precipitates ranged from 0.8~1.2 μm and were randomly distributed (Figure 3d), while the laminar precipitates with thickness of about 0.2 μm were located at grain boundaries (Figure 3c). EDS analysis indicates that the precipitates formed after HT are mainly metallic carbides, which are Cr- and Mo-enriched $M_{23}C_6$ phase in the grain interior (Figure 3d) and laminar-structured carbides with depleted Fe, Cr, and Ni at grain boundaries (Figure 3c). Different from the 1177-AC sample, water quench can effectively suppress the formation of carbides in the 1177-WQ sample, especially the formation of lamellar-structured carbides at grain boundaries, resulting in a relatively uniform structure with some annealing twins (Figure 3b). Figure 3e shows the crystal structure of L-PBFed and HTed specimens, indicating that the specimens kept a face-centered cubic (FCC) matrix before and after HT.

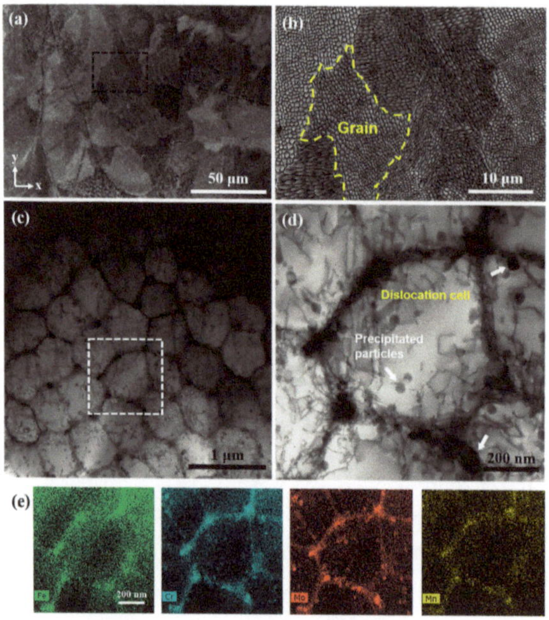

Figure 2. Microstructures of the as-fabricated GH3536 superalloy. (**a**,**b**) SEM images show the typical microstructure perpendicular to the building direction of L-PBF. (**c**) Bright-field STEM image shows the structure of dislocation cells with some precipitates. (**d**) Zoomed-in area marked by the white rectangle in (**c**). (**e**) Element-distribution maps of the selected area in (**c**), demonstrating the element segregation at the walls of dislocation cells.

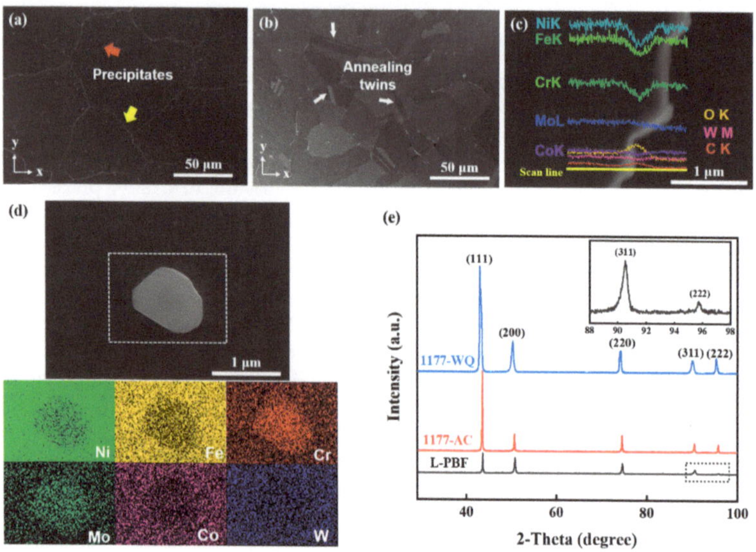

Figure 3. Microstructures of the HTed samples. (**a**,**b**) Microstructures of 1177-AC and 1177-WQ samples predencular to building direction, respectively. (**c**) The laminar precipitate of 1177-AC with a thickness of ~0.2 μm at the grain boundary and the EDS line-scan patterns (marked by yellow line). (**d**) EDS mapping of precipitates in the 1177-AC interior which are randomly distributed and ranging from 0.8~1.2 μm. (**e**) The XRD patterns of L-PBFed, 1177-AC, and 1177-WQ samples.

3.2. Mechanical Behaviors

Tensile tests were conducted at room temperature to reveal the effect of HT on the mechanical properties of L-PBFed samples. Figure 4a,b show the engineering and true stress–strain curves of as-fabricated and HTed samples, respectively. Table 2 summarizes the mechanical properties of different samples. Clearly, even with high residual stress, the as-printed samples typically have higher yield strength (667 MPa) and ultimate tensile strength (835 MPa), but a relatively low elongation as compared with that of HTed samples. After HT, the samples show decreased tensile strengths, especially the yield strength, but an enhanced elongation from 40% to 68%. The high yield strength of the L-PBFed sample mainly originates from the fine dendrites and dislocation structures according to the Hall–Patch effect [24,40]. When HTed at solution temperature, the residual stress was released and the typical characters of L-PBFed alloy disappeared by forming a homogenized microstructure with equiaxed grains [31]. The grown grains, eliminated dendrites, and relaxed dislocations could result in lower strength but much higher elongation, while the released residual stress and recrystallized grains could contribute to a higher strain-hardening ability. In the true stress–strain curve, the ultimate tensile strengths of HTed samples are even higher than those of as-printed samples due to the significant strain hardening. However, the strain-hardening rate of 1177-AC maintains a platform and is far beyond other specimens with the increase in true strain (Figure 4c). We hypothesize that due to the pinning effect of precipitates on dislocation motion, the formation of carbide precipitates on grain boundaries results in dislocation pile-up and contributes to the strain hardening. In other words, the precipitates at grain boundaries delay the necking, thus leading to a higher strain-hardening rate. However, due to the deformation incompatibility of brittle carbides, the strengthening effect of precipitates on grain boundaries should be limited. In the 1177-WQ sample, the recrystallized grains with lower precipitate density enable a more uniform elongation but lower strain-hardening capability.

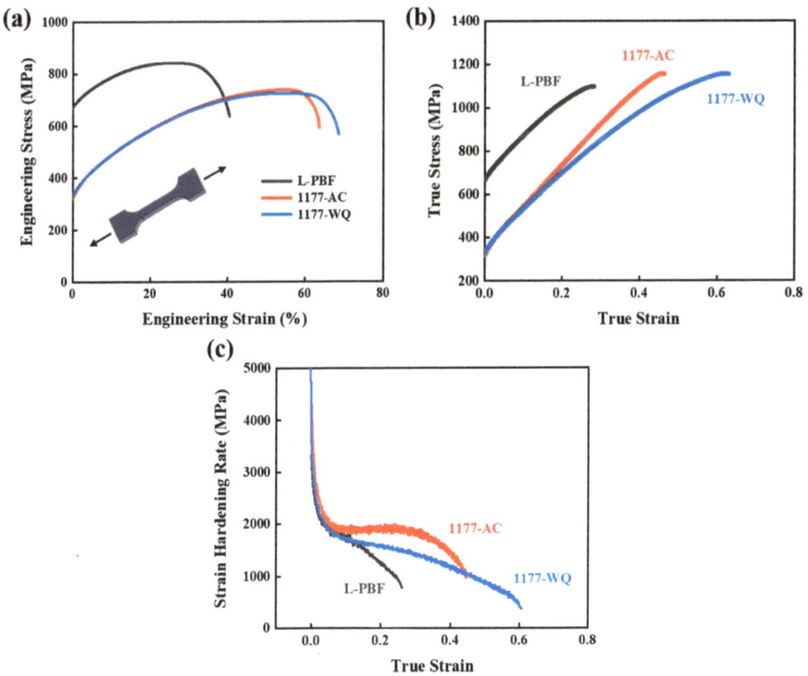

Figure 4. Mechanical properties of the L-PBFed and HTed samples. (**a**,**b**) Engineering and true stress–strain curves of different samples. (**c**) Strain-hardening rate of different samples.

Table 2. Mechanical properties of GH3536 superalloy at room temperature.

Samples	Yield Strength (MPa)	Ultimate Tensile Strength (MPa)	Elongation (%)
L-PBF	667 ± 3	835 ± 6	40.21 ± 0.74
1177-AC	323 ± 2	731 ± 5	63.01 ± 0.62
1177-WQ	334 ± 3	721 ± 3	68.70 ± 0.04

Figure 5 shows the typical fracture morphologies of different samples. The fracture surface of as-printed L-PBF samples contains several large pores at the center (Figure 5a), which should be intrinsic printing defects. Obviously, even very few of these defects could result in the generation of large pores and cracks during the tensile test. In the HTed samples, both the size and density of pores on their fracture surface are much smaller than those of the L-PBFed sample (Figure 5d,g), while the symbol of plastic deformation—dimples becomes larger, suggesting that HT can improve the ductility of the as-printed samples. Although the mechanical properties of 1177-AC and 1177-WQ samples are similar, their fracture morphologies are much different. As shown in Figure 5d, the fracture surface of the 1177-AC sample contains numerous cracks, which propagate toward the interior of the materials (Figure 5e). These cracks mainly derive from the pores and propagate zigzags, which are probably induced by the laminar-structured carbides on grain boundaries. After air cooling, the carbides precipitate out in the grain and form a continuous layer along the grain boundary. Though $M_{23}C_6$ in the grain interior can provide sites for the nucleation of dimples (Figure 5f) and increase tensile instability, the fracture of 1177-AC is mainly determined by precipitates at grain boundaries. These brittle carbide layers can act as a barrier shell to block the dislocation motion, but will induce grain-boundary fracture as the strain accumulates. With these carbides at grain boundaries, cracks possibly initiate from the inherent pores induced by the L-PBF, and then develop gradually along the brittle grain boundary. This hypothesis is confirmed by the observation of the polished fracture surface of the 1177-AC sample (Figure 6). Apparently, several transgranular cracks nucleate from or nearby a pore and then develop along the grain boundary (Figure 6a). Note that some cracks may not directly connect to the pores, but can be induced by the large deformation incompatibility at the grain boundary near a pore due to the high stress concentration. Although the crack propagation can consume some deformation energy and thus contribute to plastic deformation, the brittle grain boundary may be harmful for cyclic deformation, which thus should be avoided. In the 1177-WQ sample, elimination of the brittle carbide layer on grain boundaries largely suppresses the cracking behavior on the fracture surface, resulting in transgranular fracture (Figure 5g,h). It is worth noting that there was an agglomeration of pores in the subsurface, and the sizes of pores are larger than dimples (Figure 5h), indicating that HT could not fully eliminate the inherent pores. Few carbides are observed in the dimples of 1177-WQ (Figure 5i), contributing to the better ductility than the 1177-AC sample.

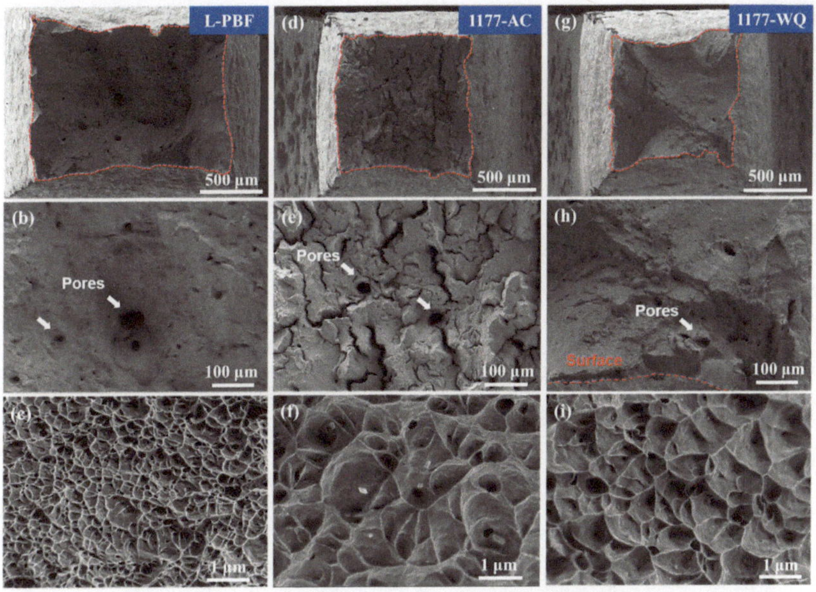

Figure 5. Fracture morphologies of different samples. (**a–c**) In L-PBF condition, (**d–f**) in 1177-AC condition, and (**g–i**) in 1177-WQ condition. The white arrows in (**b,e,h**) indicate the probable inherent pores in fracture surface. The red dotted box highlights the boundary of fracture surface.

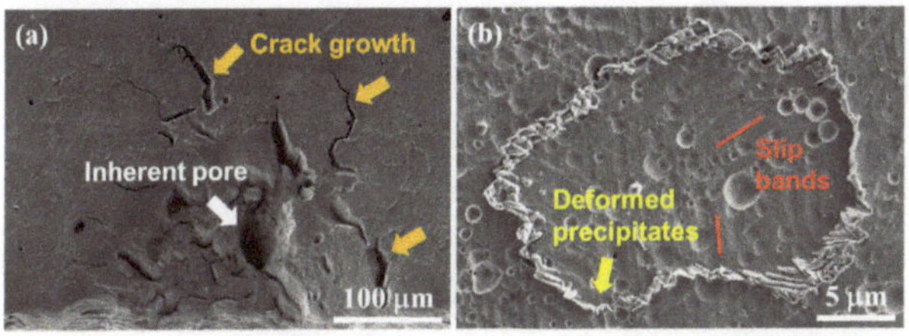

Figure 6. The polished fracture surface of the 1177-AC specimen. (**a**) The crack near an inherent pore. (**b**) The precipitated phases on the grain boundaries can act as a barrier for dislocation motion.

3.3. Structure of Deformed Samples

The high strength of the L-PBFed sample mainly originates from fine microstructure with dissatisfying low ductility limited by pores and other 3D-printing defects. In 1177-AC and 1177-WQ samples, the coarsened grains and reduced dislocation density weaken the influence of additive manufacturing defects on the mechanical properties. The homogenization and coarsened grains result in lower yield strength but higher ductility. To further understand the mechanical properties, the deformation structure of different samples was investigated. In the as-printed L-PBF sample, because of the high density of tangled dislocations, it is difficult to characterize the dislocation structure of deformed sample. In the HTed samples, the dislocation cell structure was largely destroyed, while some entangled dislocations and residual precipitates remain (Figure 7a,e), which are mainly 60° a/2 [110] full dislocations (1/6 [112] Shockley partial dislocations with stacking fault were also observed, Figure 7b,c). The previous study [41] revealed that the stacking-fault energy of GH3536 superalloy is relatively low (comparable with that of medium-/high-entropy alloys), such that full dislocations are prone to transform to extended dislocations at the early stage of plastic deformation. Since HT cannot fully eliminate the high-density dislocations produced by L-PBF, the structure of extended dislocations thus can be observed in the HTed sample. The dislocation density of the 1177-AC sample is higher than that of the 1177-WQ sample, in which tangled dislocations interweave to form dislocation braids. This might be because a large number of solid-solution elements were precipitated from the grain, and the laminar precipitates at grain boundaries may induce large internal stress interior of the grains. Some annealing twins were observed in both 1177-AC and 1177-WQ samples (Figure 7b), which can contribute to the strain hardening via dislocation–twin interaction [42,43]. Annealing twins divided the grains into smaller subspace, contributing to the Hall–Petch-type hardening. After tensile deformation, numbers of dislocations and slip bands were generated inside the 1177-AC specimen (Figure 7d). This could be ascribed to the contribution of carbides in the grain boundary that effectively hindered the dislocation motion and resulted in a dislocation pile-up inside the grain, contributing to the enhanced strain-hardening rate during the middle stage of deformation of air-cooling samples. In the 1177-WQ sample, dislocation multiplication, pile-up, and entanglement occurred abundantly in the grain interior, near the annealing twins (Figure 7e) and at the grain boundaries. The pre-existing annealing twins decreased the mean free path of dislocations during deformation, enhancing the work hardening ability (Figure 7f). It is also worth noting that precipitates cannot be fully eliminated via the HT at 1177 °C even with the water quench process (indicated by yellow arrows in Figure 7a,e). These particles can hinder the dislocation motion and contribute to the strengthening of HTed samples, but have not been fully investigated at present.

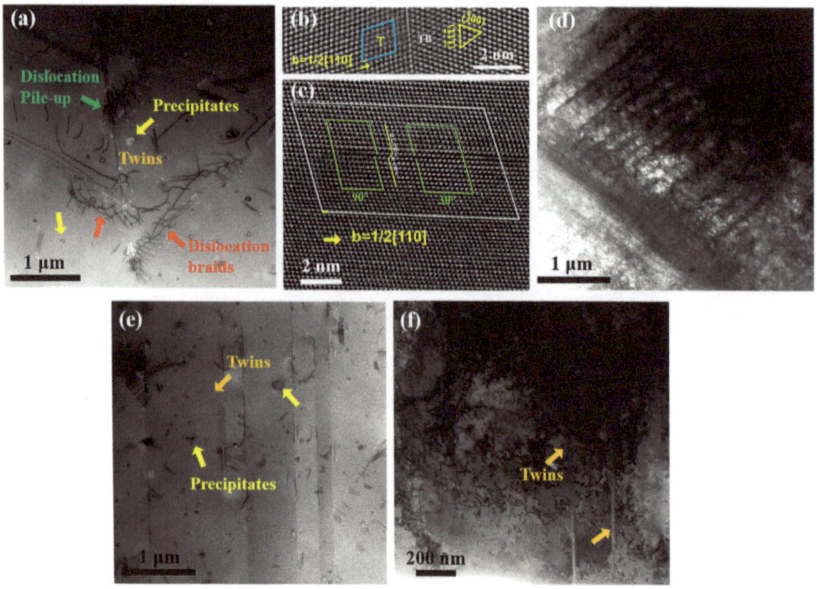

Figure 7. The STEM observation of the HTed samples. The 1177-AC before (**a**–**c**) and after (**d**) the tensile deformation. The 1177-WQ before (**e**,**f**) after the tensile deformation.

4. Discussion

4.1. Deformation of L-PBFed Sample

The microstructure of L-PBFed materials is mainly determined by the processing parameters and solidification conditions. Typically, the dislocations of L-PBFed components originate from the thermal expansion and contraction during solidification, and dislocation-cell structures are formed along the boundaries of the columnar dendrites, which maintain the same orientation as dendrites [38]. Due to the thermal cycling effect of laser-additive manufacturing, supersaturated elements are prone to precipitate from the alloy matrix and result in local segregation [39]. Hence, segregations in dislocation cells as well as their boundaries can be understood. With the dense cellular dislocation structure, the yield strength of L-PBFed materials was often described by the conventional Hall–Petch effect. The grain size d can be determined by the diameter of the dislocation cell to fully depict the high strength of the L-PBFed sample [24,40]. However, the temperature gradient between the melting-pool boundary and its upper surface along the building direction often causes insufficient melting in the metal part [17,44,45]. These areas can significantly decrease the ductility of L-PBFed specimens, due to the deformation-induced stress concentration and high possibility of inherent pores at the melting-pool boundaries, as demonstrated by our experimental observations in Figures 4a and 5a,b.

4.2. Deformation of HTed Sample

HT at solution temperature can dramatically change the microstructural characteristics of the L-PBFed specimen (Figure 3a,b). Coarsened grain and homogenized microstructure after HT were favorable to enhance the tensile stability and ductility of L-PBFed samples. However, dissolved dendrites and reduced dislocation density and cell structures would decrease the tensile strength of HTed samples (Figure 4a). For HTed samples, the grain size of 1177-AC (~84 μm) is slightly larger than 1177-WQ (~65 μm). However, as shown in Figure 4, 1177-AC possesses a slightly lower yield strength, higher strain-hardening rate, and lower elongation than 1177-WQ, indicating that the carbide precipitates at grain boundaries introduced by air cooling significantly changed the mechanical behaviors of the HTed sample and decreased the tensile stability, especially the ductility of 1177-AC. Furthermore, the XRD patterns in Figure 3e show that 1177-AC possesses a lower concentration of solid-solution elements than 1177-WQ (with the average lattice parameters of L-PBF, 1177-AC, 1177-WQ about 3.5915 Å, 3.5951 Å, and 3.6190 Å, respectively), demonstrating that lots of solid-solution elements (Cr and Mo) precipitated during the air-cooling process and contributed to a lower tensile strength. However, the ultimate tensile strength of 1177-AC is slightly higher than 1177-WQ, although 1177-AC possessed larger grain size and fewer solid-solution-strengthening elements in matrix. These results further indicate that the carbide precipitates at grain boundaries can contribute to the continuing strain hardening of 1177-AC.

For the solid-solution alloys, the Taylor hardening law [41,46] was used to describe the interactions of dislocations with increasing strain rates:

$$\sigma_T = \alpha G b \sqrt{\rho} \tag{1}$$

where α is constant, G is the shear modulus, b is the Burgers vector, and ρ is the dislocation density. The strain-hardening behaviors of 1177-AC and 1177-WQ samples are discussed according to Equation (1). At the initial stage of plastic deformation, dislocations are activated and then rapidly proliferated combining with their pileups at grain boundaries, which in turn further hinder the motion of other dislocations. As the strain increases, the density of dislocations continuously increases. As shown in Figure 7f, dislocations in the 1177-WQ sample form a complex structure through dislocation–dislocation and dislocation–twin interactions. In the 1177-AC sample, as dislocations slip to grain boundaries, the shell-like carbides segregated at grain boundaries can greatly hinder dislocation motion, by which, as the strain accumulates, high deformation incompatibility occurs, resulting in cracks along grain boundaries (Figure 5e). This deformation process is illustrated by the observation of high-density slip bands in the interior of the grains and the transgranular cracks in Figure 6a.

Figure 8 shows the enhanced ductility and acceptable ultimate tensile strength of HTed GH3536 superalloy compared with Hastalloy X in other literatures, including L-PBFed/HTed samples and wrought samples. However, numerous built-in pores were still generated during plastic deformation and can be seen on the fracture surface of HTed samples (Figure 5d,g). This means that HTed samples cannot fully eliminate the inherent defects induced by L-PBF, which can impact the crack nucleation and fracture of HTed samples. Hot isostatic pressing (HIP) is often conducted to reduce the impact of these

L-PBF-induced defects, during which carbides may grow during the thermal exposure of HIP and influence the ductility of materials [30].

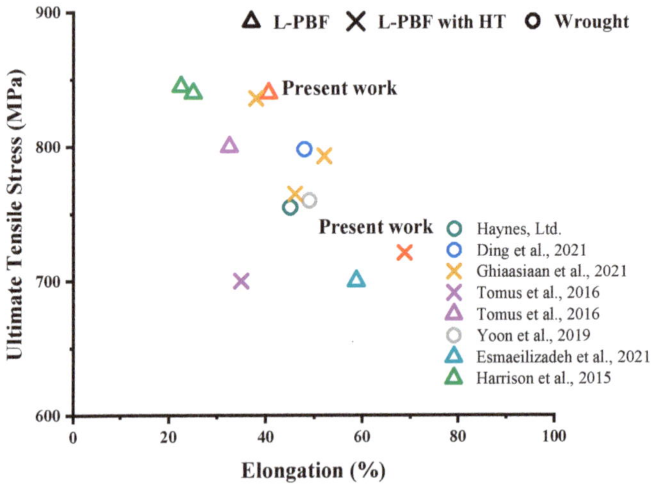

Figure 8. The comparison of ultimate tensile strength and ductility of GH3536 superalloy from our work and literatures [18,24,30,34,41,47,48].

5. Conclusions

We fabricated the GH3536 alloys via L-PBF methods and studied their microstructure and mechanical properties under different conditions. Post-heat-treatments reveal that different cooling processes result in totally different microstructure and fracture morphology in HTed GH3536 superalloy. The main results can be concluded below:

(a) A near full-dense specimen containing a high density of fine dendrites and cellular dislocations was fabricated by L-PBF, which showed high yield strength and ultimate tensile strength, but relative lower ductility.
(b) The longer homogenization time and air-cooling process lead to coarse grains but laminar precipitates at grain boundaries. The carbide precipitates may act as stable barriers that hinder the dislocation motion for strength, resulting in a higher strain-hardening rate but lower ductility.
(c) Constraining the formation of precipitates in grain boundaries by water quench is necessary for HTed samples, which contribute to an enhanced ductility and acceptable strength compared with wrought samples.

Author Contributions: Conceptualization, H.N., Q.Z. and J.W.; methodology, H.N. and J.W.; validation, K.Z. and J.W.; formal analysis, H.N. and J.W.; investigation, H.N. and Q.Z.; resources, Q.Z., K.Z. and J.W.; writing—original draft preparation, H.N.; writing—review and editing, H.N., Y.C. and J.W.; visualization, H.N.; project administration, J.W.; funding acquisition, J.W. All authors have read and agreed to the published version of the manuscript.

Funding: The authors gratefully acknowledge the support of the Aero Engine Corporation of China (Grant HFZL2019CXY001). J.W. acknowledges the support of the National Natural Science Foundation of China (Grant 52071284).

Institutional Review Board Statement: Not applicable.

Informed Consent Statement: Not applicable.

Data Availability Statement: Not applicable.

Conflicts of Interest: The authors declare no conflict of interest.

References

1. Bochenek, K.; Węglewski, W.; Strojny-Nędza, A.; Pietrzak, K.; Chmielewski, T.; Chmielewski, M.; Basista, M. Microstructure, Mechanical, and Wear Properties of NiCr-Re-Al_2O_3 Coatings Deposited by HVOF, Atmospheric Plasma Spraying, and Laser Cladding. *J. Therm. Spray Technol.* **2022**, *31*, 1609–1633. [CrossRef]
2. Kolodziejczak, P.; Golanski, D.; Chmielewski, T.; Chmielewski, M. Microstructure of Rhenium Doped Ni-Cr Deposits Produced by Laser Cladding. *Materials* **2021**, *14*, 2745. [CrossRef] [PubMed]
3. Zhang, L.; Song, J.; Wu, W.; Gao, Z.; He, B.; Ni, X.; Long, Q.; Lu, L.; Zhu, G. Effect of processing parameters on thermal behavior and related density in GH3536 alloy manufactured by selective laser melting. *J. Mater. Res.* **2019**, *34*, 1405–1414. [CrossRef]
4. DebRoy, T.; Wei, H.L.; Zuback, J.S.; Mukherjee, T.; Elmer, J.W.; Milewski, J.O.; Beese, A.M.; De Wilson-Heid, A.; Zhang, W. Additive manufacturing of metallic components—Process, structure and properties. *Prog. Mater. Sci.* **2018**, *92*, 112–224. [CrossRef]
5. Kim, I.S.; Choi, B.G.; Jung, J.E.; Do, J.; Jo, C.Y. Effect of microstructural characteristics on the low cycle fatigue behaviors of cast Ni-base superalloys. *Mater. Charact.* **2015**, *106*, 375–381. [CrossRef]
6. Xiao, G.; Jiang, J.; Wang, Y.; Liu, Y.; Zhang, Y.; Tian, Y. Effects of Isothermal Temperature and Soaking Time on Water Quenched Microstructure of Nickel-Based Superalloy GH3536 Semi-Solid Billets. *Materials* **2021**, *14*, 4668. [CrossRef]
7. Berman, B. 3-D printing: The new industrial revolution. *Bus. Horiz.* **2012**, *55*, 155–162. [CrossRef]
8. Mumtaz, K.A.; Erasenthiran, P.; Hopkinson, N. High density selective laser melting of Waspaloy®. *J. Mater. Processing Technol.* **2008**, *195*, 77–87. [CrossRef]
9. Yin, Y.; Tan, Q.; Bermingham, M.; Mo, N.; Zhang, J.; Zhang, M.-X. Laser additive manufacturing of steels. *Int. Mater. Rev.* **2021**, *67*, 487–573. [CrossRef]
10. Shunmugavel, M.; Polishetty, A.; Littlefair, G. Microstructure and Mechanical Properties of Wrought and Additive Manufactured Ti-6Al-4V Cylindrical Bars. *Procedia Technol.* **2015**, *20*, 231–236. [CrossRef]
11. Zhao, J.-R.; Hung, F.-Y.; Lu, C.-S.; Lai, I.C. Comparison of Laser Powder Bed Fusion and Cast Inconel 713 Alloy in Terms of Their Microstructure, Mechanical Properties, and Fatigue Life. *Adv. Eng. Mater.* **2021**, *23*, 2001366. [CrossRef]
12. Manvatkar, V.; De, A.; DebRoy, T. Heat transfer and material flow during laser assisted multi-layer additive manufacturing. *J. Appl. Phys.* **2014**, *116*, 124905. [CrossRef]
13. Thijs, L.; Kempen, K.; Kruth, J.-P.; Van Humbeeck, J. Fine-structured aluminium products with controllable texture by selective laser melting of pre-alloyed AlSi10Mg powder. *Acta Mater.* **2013**, *61*, 1809–1819. [CrossRef]
14. Carlton, H.D.; Haboub, A.; Gallegos, G.F.; Parkinson, D.Y.; MacDowell, A.A. Damage evolution and failure mechanisms in additively manufactured stainless steel. *Mater. Sci. Eng. A* **2016**, *651*, 406–414. [CrossRef]
15. Niendorf, T.; Leuders, S.; Riemer, A.; Richard, H.A.; Tröster, T.; Schwarze, D. Highly Anisotropic Steel Processed by Selective Laser Melting. *Metall. Mater. Trans. B* **2013**, *44*, 794–796. [CrossRef]
16. Wen, S.; Li, S.; Wei, Q.; Yan, C.; Zhang, S.; Shi, Y. Effect of molten pool boundaries on the mechanical properties of selective laser melting parts. *J. Mater. Processing Technol.* **2014**, *214*, 2660–2667. [CrossRef]
17. Khairallah Saad, A.; Martin Aiden, A.; Lee Jonathan, R.I.; Guss, G.; Calta Nicholas, P.; Hammons Joshua, A.; Nielsen Michael, H.; Chaput, K.; Schwalbach, E.; Shah Megna, N.; et al. Controlling interdependent meso-nanosecond dynamics and defect generation in metal 3D printing. *Science* **2020**, *368*, 660–665. [CrossRef]
18. Esmaeilizadeh, R.; Keshavarzkermani, A.; Ali, U.; Behravesh, B.; Bonakdar, A.; Jahed, H.; Toyserkani, E. On the effect of laser powder-bed fusion process parameters on quasi-static and fatigue behaviour of Hastelloy X: A microstructure/defect interaction study. *Addit. Manuf.* **2021**, *38*, 101805. [CrossRef]
19. Li, R.; Liu, J.; Shi, Y.; Wang, L.; Jiang, W. Balling behavior of stainless steel and nickel powder during selective laser melting process. *Int. J. Adv. Manuf. Technol.* **2011**, *59*, 1025–1035. [CrossRef]
20. Ali, H.; Ghadbeigi, H.; Mumtaz, K. Effect of scanning strategies on residual stress and mechanical properties of Selective Laser Melted Ti6Al4V. *Mater. Sci. Eng. A* **2018**, *712*, 175–187. [CrossRef]
21. Al Mangour, B.; Grzesiak, D.; Yang, J.-M. Scanning strategies for texture and anisotropy tailoring during selective laser melting of TiC/316L stainless steel nanocomposites. *J. Alloys Compd.* **2017**, *728*, 424–435. [CrossRef]
22. Wang, N.; Mokadem, S.; Rappaz, M.; Kurz, W. Solidification cracking of superalloy single- and bi-crystals. *Acta Mater.* **2004**, *52*, 3173–3182. [CrossRef]
23. Ojo, O.A.; Richards, N.L.; Chaturvedi, M.C. Liquation of various phases in HAZ during welding of cast Inconel* 738LC. *Mater. Sci. Technol.* **2013**, *20*, 1027–1034. [CrossRef]
24. Harrison, N.J.; Todd, I.; Mumtaz, K. Reduction of micro-cracking in nickel superalloys processed by Selective Laser Melting: A fundamental alloy design approach. *Acta Mater.* **2015**, *94*, 59–68. [CrossRef]
25. Tomus, D.; Jarvis, T.; Wu, X.; Mei, J.; Rometsch, P.; Herny, E.; Rideau, J.F.; Vaillant, S. Controlling the Microstructure of Hastelloy-X Components Manufactured by Selective Laser Melting. *Phys. Procedia* **2013**, *41*, 823–827. [CrossRef]
26. Wang, F.; Wu, X.H.; Clark, D. On direct laser deposited Hastelloy X: Dimension, surface finish, microstructure and mechanical properties. *Mater. Sci. Technol.* **2014**, *27*, 344–356. [CrossRef]
27. H[illegible]atzadeh, S.M.H.; Parab, N.D.; Guo, Q.; Qu, M.; Xiong, L.; Zhao, C.; Escano, L.I.; Fezzaa, K.; Everhart, W.; Sun, T.; et al. Direct [illegible]vation of pore formation mechanisms during LPBF additive manufacturing process and high energy density laser welding. [illegible] *J. Mach. Tools Manuf.* **2020**, *153*, 103555. [CrossRef]

28. Chauvet, E.; Kontis, P.; Jägle, E.A.; Gault, B.; Raabe, D.; Tassin, C.; Blandin, J.-J.; Dendievel, R.; Vayre, B.; Abed, S.; et al. Hot cracking mechanism affecting a non-weldable Ni-based superalloy produced by selective electron Beam Melting. *Acta Mater.* **2018**, *142*, 82–94. [CrossRef]
29. Sanchez-Mata, O.; Muñiz-Lerma, J.A.; Wang, X.; Atabay, S.E.; Attarian Shandiz, M.; Brochu, M. Microstructure and mechanical properties at room and elevated temperature of crack-free Hastelloy X fabricated by laser powder bed fusion. *Mater. Sci. Eng. A* **2020**, *780*, 139177. [CrossRef]
30. Tomus, D.; Tian, Y.; Rometsch, P.A.; Heilmaier, M.; Wu, X. Influence of post heat treatments on anisotropy of mechanical behaviour and microstructure of Hastelloy-X parts produced by selective laser melting. *Mater. Sci. Eng. A* **2016**, *667*, 42–53. [CrossRef]
31. Marchese, G.; Lorusso, M.; Parizia, S.; Bassini, E.; Lee, J.-W.; Calignano, F.; Manfredi, D.; Terner, M.; Hong, H.-U.; Ugues, D.; et al. Influence of heat treatments on microstructure evolution and mechanical properties of Inconel 625 processed by laser powder bed fusion. *Mater. Sci. Eng. A* **2018**, *729*, 64–75. [CrossRef]
32. Cunningham, R.; Nicolas, A.; Madsen, J.; Fodran, E.; Anagnostou, E.; Sangid, M.D.; Rollett, A.D. Analyzing the effects of powder and post-processing on porosity and properties of electron beam melted Ti-6Al-4V. *Mater. Res. Lett.* **2017**, *5*, 516–525. [CrossRef]
33. Tammas-Williams, S.; Withers, P.J.; Todd, I.; Prangnell, P.B. Porosity regrowth during heat treatment of hot isostatically pressed additively manufactured titanium components. *Scr. Mater.* **2016**, *122*, 72–76. [CrossRef]
34. Ghiaasiaan, R.; Muhammad, M.; Gradl, P.R.; Shao, S.; Shamsaei, N. Superior tensile properties of Hastelloy X enabled by additive manufacturing. *Mater. Res. Lett.* **2021**, *9*, 308–314. [CrossRef]
35. Sun, S.; Teng, Q.; Xie, Y.; Liu, T.; Ma, R.; Bai, J.; Cai, C.; Wei, Q. Two-step heat treatment for laser powder bed fusion of a nickel-based superalloy with simultaneously enhanced tensile strength and ductility. *Addit. Manuf.* **2021**, *46*, 102168. [CrossRef]
36. Chen, Z.; Wu, X.; Tomus, D.; Davies, C.H.J. Surface roughness of Selective Laser Melted Ti-6Al-4V alloy components. *Addit. Manuf.* **2018**, *21*, 91–103. [CrossRef]
37. Cherry, J.A.; Davies, H.M.; Mehmood, S.; Lavery, N.P.; Brown, S.G.R.; Sienz, J. Investigation into the effect of process parameters on microstructural and physical properties of 316L stainless steel parts by selective laser melting. *Int. J. Adv. Manuf. Technol.* **2014**, *76*, 869–879. [CrossRef]
38. Bertsch, K.M.; Meric de Bellefon, G.; Kuehl, B.; Thoma, D.J. Origin of dislocation structures in an additively manufactured austenitic stainless steel 316L. *Acta Mater.* **2020**, *199*, 19–33. [CrossRef]
39. Li, X.P.; Ji, G.; Chen, Z.; Addad, A.; Wu, Y.; Wang, H.W.; Vleugels, J.; Van Humbeeck, J.; Kruth, J.P. Selective laser melting of nano-TiB2 decorated AlSi10Mg alloy with high fracture strength and ductility. *Acta Mater.* **2017**, *129*, 183–193. [CrossRef]
40. Zhong, Y.; Liu, L.; Wikman, S.; Cui, D.; Shen, Z. Intragranular cellular segregation network structure strengthening 316L stainless steel prepared by selective laser melting. *J. Nucl. Mater.* **2016**, *470*, 170–178. [CrossRef]
41. Ding, Q.; Bei, H.; Wei, X.; Gao, Y.F.; Zhang, Z. Nano-twin-induced exceptionally superior cryogenic mechanical properties of a Ni-based GH3536 (Hastelloy X) superalloy. *Mater. Today Nano* **2021**, *14*, 100110. [CrossRef]
42. Zhu, Q.; Zhao, S.C.; Deng, C.; An, X.H.; Song, K.X.; Mao, S.X.; Wang, J.W. In situ atomistic observation of grain boundary migration subjected to defect interaction. *Acta Mater.* **2020**, *199*, 42–52. [CrossRef]
43. Wang, J.; Cao, G.; Zhang, Z.; Sansoz, F. Size-dependent dislocation-twin interactions. *Nanoscale* **2019**, *11*, 12672–12679. [CrossRef] [PubMed]
44. Khairallah, S.A.; Anderson, A.T.; Rubenchik, A.; King, W.E. Laser powder-bed fusion additive manufacturing: Physics of complex melt flow and formation mechanisms of pores, spatter, and denudation zones. *Acta Mater.* **2016**, *108*, 36–45. [CrossRef]
45. Wang, T.; Zhu, Y.Y.; Zhang, S.Q.; Tang, H.B.; Wang, H.M. Grain morphology evolution behavior of titanium alloy components during laser melting deposition additive manufacturing. *J. Alloys Compd.* **2015**, *632*, 505–513. [CrossRef]
46. Voyiadjis, G.Z.; Yaghoobi, M. Introduction: Size effects in materials. In *Size Effects in Plasticity*; Elsevier: Amsterdam, The Netherlands, 2019; pp. 1–79.
47. Available online: http://www.haynes.ch/doc/HASTELLOY_X.pdf (accessed on 27 June 2022).
48. Yoon, D.; Heo, I.; Kim, J.; Chang, S.; Chang, S. Hold Time-Low Cycle Fatigue Behavior of Nickel Based Hastelloy X at Elevated Temperatures. *Int. J. Precis. Eng. Manuf.* **2019**, *20*, 147–157. [CrossRef]

metals

Article

Influence of Replacing Molybdenum with Tungsten on the Creep Fracture Property of Waspaloy Nickel-Based Alloy

Hanxin Yao [1,2,*], Jianxin Dong [1,*], Zhihua Gong [3,4], Jiqing Zhao [4] and Gang Yang [4]

1 School of Materials Science and Engineering, University of Science and Technology Beijing, Beijing 100083, China
2 Beijing Mechanical & Electronical Technology Research Institute, Beijing 100083, China
3 School of Materials and Metallurgy, Inner Mongolia University of Science & Technology, Baotou 014010, China
4 Institute for Special Steels, Central Iron and Steel Research Institute, Beijing 100081, China
* Correspondence: postustb@163.com (H.Y.); jxdong@ustb.edu.cn (J.D.)

Abstract: Alloys meeting the requirements of "700 °C and above" advanced ultra-super-critical technology, with higher thermal efficiency, have been developed in recent years. Here, a new wrought Ni-based superalloy with excellent high-temperature creep strength based on Waspaloy has been developed and is proposed as a candidate material for application in 700 °C class advanced ultra-super-critical steam turbine blades. In this new alloy, the Molybdenum (Mo) in Waspaloy is partially replaced by Tungsten (W). Creep tests have shown that this new Ni-based alloy has a 70 MPa higher creep-rupture strength than that of Waspaloy at 700 °C by extrapolating the experimental data. Detailed creep-rupture mechanisms have been analyzed by means of scanning electron microscopy, transmission electron microscopy, and chemical phase analysis with a view to devising potential approaches for performance improvements. The results showed that the partial replacement of Mo by W had negligible effect on the composition of carbides precipitated in the alloy. Instead, the amount of the γ' phase was significantly increased, and mismatch between the γ and γ' phases was reduced. In this way, the stability of the γ' phase was increased, its coarsening rate was reduced, and its critical shear stress was increased. As a result, the high-temperature creep-fracture strength of the new alloy was increased.

Keywords: advanced ultra-super-critical; Waspaloy; tungsten; creep-fracture strength; critical shear stress

1. Introduction

In recent years, many countries, including China, have begun to focus on the research and development of advanced ultra-super-critical technology with a steam temperature of 700 °C, whereby the thermal efficiency of power stations is expected to break through the 50% threshold. The extremely harsh service conditions of this grade of power station have led to the development of specifically designed superalloys as candidate materials [1–3].

As the core components of turbine blades, the main candidate materials in various countries are Waspaloy, Nimonic105, USC141, M252, and so on [4,5]. Waspaloy (Ni-20Cr-13Co-4Mo-1.4Al-3Ti-B-Zr), a precipitation-hardening nickel-based superalloy, was developed by Pratt and Whitney in the United States in the 1950s [6] for utilization in aviation engines and power machines. Such alloys are now essential materials for manufacturing turbine disks and blades, showing good tensile and endurance properties below 760 °C and good oxidation resistance above 870 °C. An advantage of Waspaloy is that it has good strength and toughness matching [7].

Waspaloy is mainly used in aero-engine turbine blade materials. M. Razumovskii et al. [8]. found that increasing the mass fraction of W would greatly improve the melting temperature of the alloy, and the creep-fracture strength would be more advantageous. Alloying element W can improve the heat resistance of the alloy, but also has the properties of being cheap, very cost-effective [9]. The Mo content increases the strength and durability of Waspaloy by solid-solution strengthening. A new type of Ni-based alloy has been designed by our research group by partially replacing Mo in Waspaloy with W. The previous study showed that the type of precipitated phases in Waspaloy alloy would not change, and the alloy maintained good structure stability during long-term aging when adding 2% W (mass fraction) to the alloy. Meanwhile, W could refine the size and reduce the growth rate of γ' phase, and improved the tensile strength of the alloy [10–13].

The influence of W on the microstructural stability and properties of such alloys has yet to be fully characterized. Therefore, in this study, we have monitored the effects of long-term aging on carbide separation, precipitation-strengthening phase formation, and the properties of the new alloy. Furthermore, the strengthening mechanism of W substitution of Mo in Waspaloy alloy is discussed.

2. Experimental Material and Procedures

The chemical composition of the test alloy in Table 1 was determined by thermodynamic calculation and combining with the alloys with different W content in the previous study [9,11]. The new alloy (designated as 2Mo2W) was obtained by replacing Mo in Waspaloy with an equal mass of W. The test alloy was melted by a "double-vacuum" process (vacuum induction smelting + vacuum arc remelting), poured into 50 kg ingots, and subsequently forged into a rod of diameter 18 mm. The forging temperature was 1150 °C, the final forging temperature was not lower than 950 °C, and the cooling method is heap cooling. The heat-treatment process involved the following sequential steps: 1080 °C for 4 h (oil cooling), 845 °C for 24 h (air cooling), and 760 °C for 16 h (air cooling). Thereafter, alloy specimens were exposed to 700 °C for 1000–10,000 h or 750 °C for 1000–5000 h in a resistance furnace, then air-cooled to room temperature.

Table 1. Mass percentage compositions of alloys (wt%).

Alloy	C	Cr	Co	Al	Ti	W	Mo	Zr	B	Ni
2Mo2W	0.04	19.49	13.56	1.44	2.95	1.97	2.19	0.028	0.01	Bal.
Waspaloy [14,15]	0.056	19.64	13.68	1.40	3.18	-	4.5	0.044	0.0063	Bal.

The above samples were cleaned with a mixture of $CuCl_2$ (10 g) + HCl (100 mL) + CH_3CH_2OH (100 mL) or electrolytically etched with HNO_3 (40%) + H_2O (60%) solution. This allowed the microstructures of carbides and the γ' phase to be observed by scanning electron microscopy (SEM, Hitachi S4300, Tokyo, Japan). Qualitative and quantitative analyses of carbides and γ' phase particles were carried out by X-ray diffraction (XRD, APD-10, Milan, Italy; employing Co-K_{α} radiation and operated at 30 kV and 25 mA) and chemical phase analysis methods. Equilibrium precipitations in the alloys were investigated by thermodynamic calculations using the Ni-9 database.

Two samples were used for each observation, three samples were used for each performance test, performance is averaged over tests.

The creep-fracture strength was tested on the CMT-105 (100 KN) creep test machine, and the samples were Φ5 mm standard high temperature tensile sample (GB6395-86). The test temperature was 700 °C, and the stresses were 550 MPa, 500 MPa, 475 MPa, 450 MPa, and 425 MPa.

Figure 1 shows the metallographic and scanning microstructure of 2Mo2W and Waspaloy alloys before long-term exposure. The difference of microstructure between the two alloys is not obvious. There are some $M_{23}C_6$ carbides distributed on the grain boundary. The large precipitations, precipitated in the grain, are unmelted primary MC carbides.

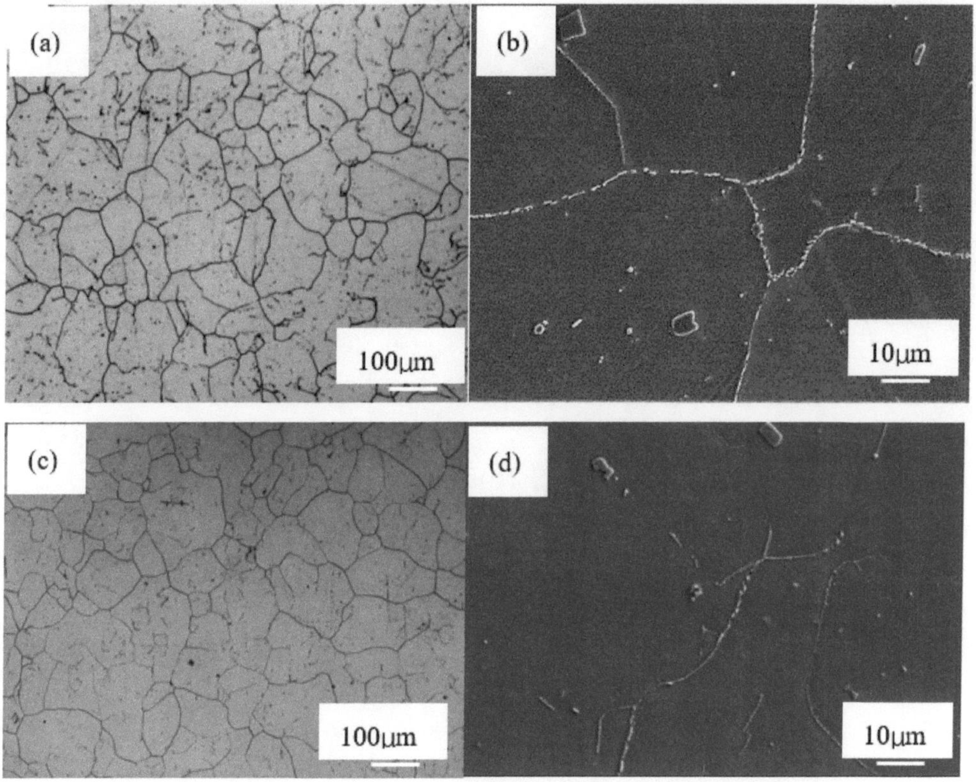

Figure 1. Microstructures of test alloys before long-term exposure. (**a**,**b**) Waspaloy; (**c**,**d**) 2Mo2W.

3. Results and Discussion

3.1. Characteristics of the Precipitation after Long-Term Exposure

(1) Evolution of carbides

Thermal-CALC thermodynamic software was applied to calculate the equilibrium precipitate composition of the new alloy containing 2% W and 2% Mo, as shown in Figure 2. The precipitate comprised small amounts of the $M_{23}C_6$ phase, MC phase, and μ-phase and a large amount of the γ' phase. Figure 2b shows an enlarged image of the pertinent region of Figure 2a, in which it can be seen more clearly that the $M_{23}C_6$ phase precipitated at 920 °C and then remained almost unchanged with increasing temperature. The μ-phase content increased in a parabolic manner as the sample was cooled, while the γ' phase content increased rapidly.

Thermodynamic calculations can only explain the theoretical equilibrium state, and the actual precipitation process will differ. Figure 3 shows the XRD patterns of the precipitated phase in the alloy before and after long-term aging. From inspection of Figure 3, it can be seen that the carbides in the alloy were $M_{23}C_6$ and TiC both before and after aging. This indicates that long-term aging at 700 °C or 750 °C has no significant effect on the type of carbides precipitated.

No precipitation of σ, τ, or other harmful phases was observed, indicating that their contents were insignificantly small and any effect on the strength of the alloy could be ignored.

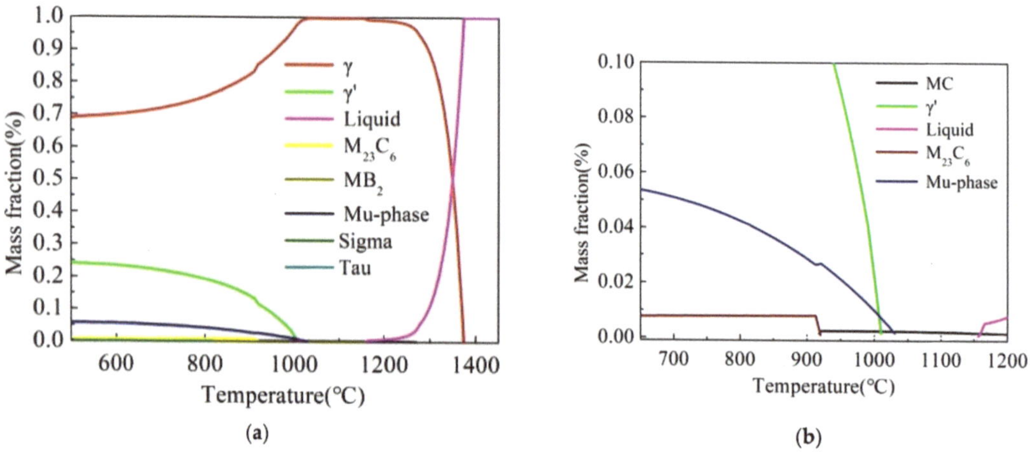

Figure 2. (**a**). Equilibrium phase diagram of the precipitate in the new alloy; (**b**) expansion of the pertinent central region in (**a**).

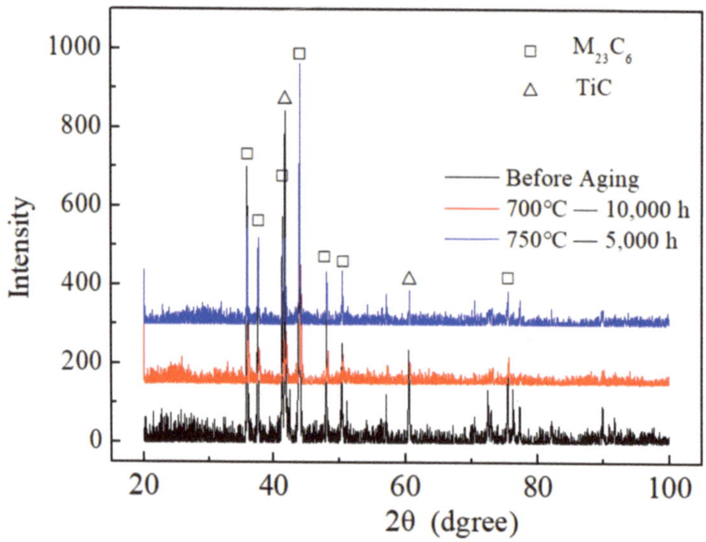

Figure 3. XRD patterns of carbides precipitated in the new alloy before and after long-term aging at different temperatures.

The variations in the contents of $M_{23}C_6$ and MC in the experimental alloy with aging time at 700 °C are shown in Figure 4. The $M_{23}C_6$ content increased steadily with aging time from 0 to 3000 h, reached a maximum at around 5000 h, and thereafter remained essentially unchanged. The MC content slowly increased with aging time from 0 to 3000 h, and thereafter showed a slight decrease.

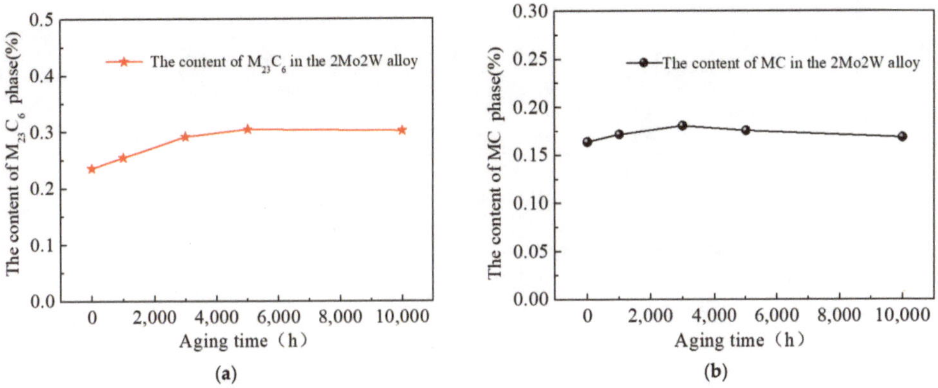

Figure 4. Temporal evolution of the contents of (**a**) $M_{23}C_6$ and (**b**) TiC during long-term treatment at 700 °C.

TiC has a high melting point and is usually stable in the range 700–750 °C [16], but during long service of Ni-based alloys, the MC phase will degenerate according to $MC + \gamma \rightarrow M_{23}C_6 + \gamma'$ or $MC + \gamma \rightarrow M_6C + \gamma'$. In some high-W alloys, the $MC + \gamma \rightarrow M_{23}C_6 + \gamma'$ degradation reaction will become $MC + \gamma \rightarrow M_{23}C_6 + \eta$ [17,18]. The η and M_6C phases are not stable and usually have adverse effects on the creep-fracture strengths of alloys.

The MC phase in the long-term aged alloy and its associated precipitation were analyzed by energy-dispersive spectrometry (EDS), and the results are shown in Figure 5 and Table 2.

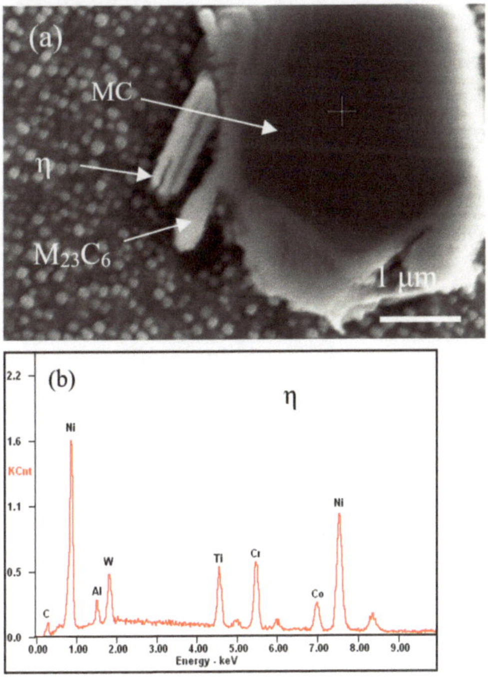

Figure 5. (**a**) Morphologies of the MC, $M_{23}C_6$ and η phases; (**b**) EDS of the η phase.

Table 2. Mass percentage compositions of precipitations (wt%).

Precipitation	C K	AlK	WL	TiK	CrK	CoK	NiK	MoL
η	0.04	1.6	1.86	7.17	12.62	11.99	64.72	-
$M_{23}C_6$	2.97	1.49	1.66	3.17	19.03	12.15	56.26	3.27
MC	1.12	0.49	12.35	50.59	8.90	4.65	16.25	6.64

Around the large TiC particles, there are two precipitated phases of two components. Compositional analysis showed one to be the $M_{23}C_6$ phase and the other to be the η phase, rich in Ti [19]. It has been shown previously that, during degradation of the MC phase with high Ti content, the η phase is precipitated. Thus, it can be surmised that the MC phase was degraded here.

(2) Evolution and coarsening kinetics of the γ' phase

Figure 6 shows high-resolution Transmission Electron Microscopy (TEM) images of the interface structure between the γ and γ' phases in the new alloy. From the diffraction pattern in Figure 6b, it can be seen that the γ' particles and the matrix in Figure 6a are both [110] zone axis. Moreover, it can be seen from the enlargement in Figure 6d of the area within the rectangle in Figure 6c that the interface between the γ and γ' phases is a perfect co-lattice, with parallel stripes and no defects such as distortions or dislocations.

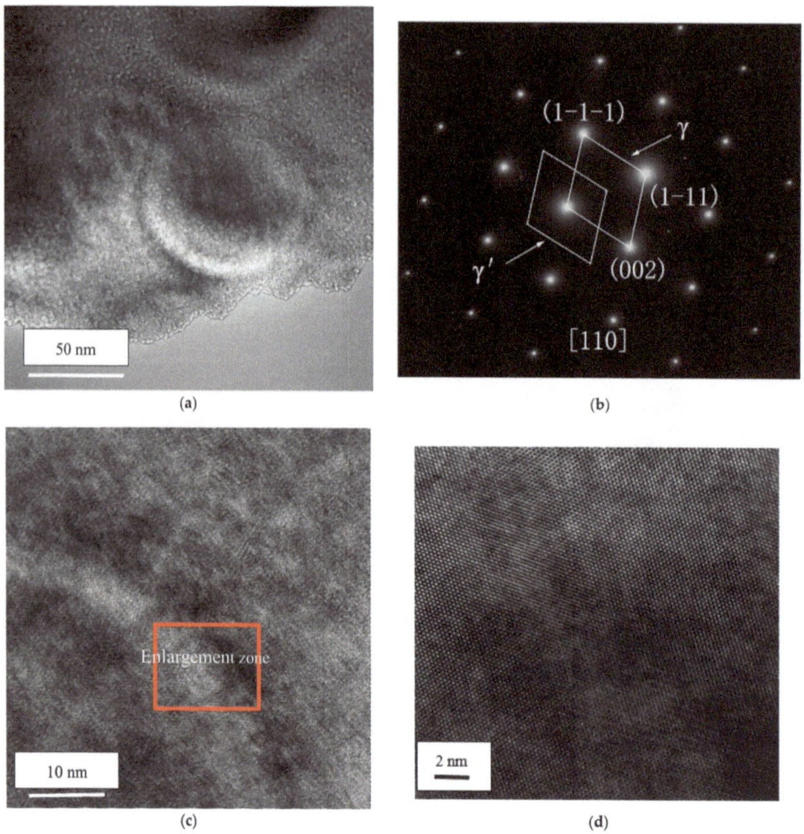

Figure 6. Interface structure between the γ and γ' phases observed by HR-TEM. (**a**) γ' phase; (**b**) selected area electron diffraction of γ'; (**c**,**d**) are HR-TEM pictures.

Ostwald ripening coarsening follows the diffusion-controlled particle growth rule: $r^3 \propto t$, where r is the particle size and t is time. In Ni-based superalloys, the coarsening rate can be written as $r^3 - r_0^3 = kt$, where r is the average size of γ' after long-term exposure, r_0 is the average γ' particle size before exposure, and k is the linear coefficient of a plot of $r^3 - r_0^3$ versus exposure time [20,21].

Table 3 shows the size of the γ' phase of test alloys after long-time aging. Figure 7 compares the coarsening rates of the γ' phase in 2Mo2W alloy and Waspaloy over 10,000 h at 700 °C. The coarsening rate of the γ' phase in Waspaloy is k = 5.76, more than twice that of the 2Mo2W alloy (k = 2.47).

Table 3. γ' phase size of test alloys before and after treatment.

Test Alloy (r/nm)	Before Aging	1000 h	2000 h	5000 h	10,000 h
2Mo2W	30.75	32.15	-	28.90	34.50
Waspaloy	31.40	-	31.80	33.50	41.35

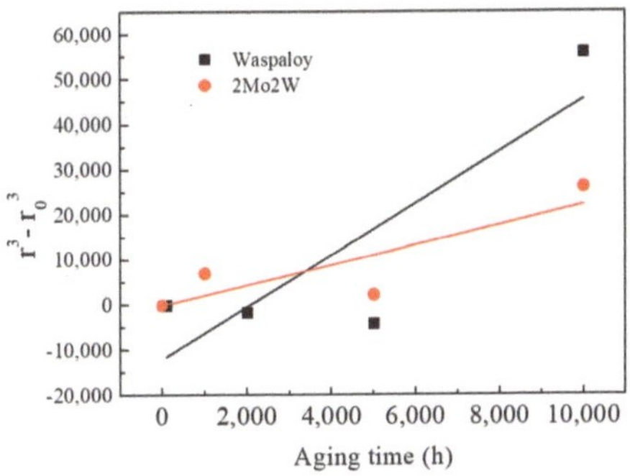

Figure 7. Coarsening rates of 2Mo2W alloy and Waspaloy during long-term aging at 700 °C.

3.2. Discussion on the Mechanism of Creep-Fracture Reinforcement

Figure 8 compares the stress–time enduring logarithmic plots for Waspaloy and the 2Mo2W alloy. Waspaloy has a durable strength of about 340 MPa at 10,000 h, around 70 MPa lower than that of the new alloy.

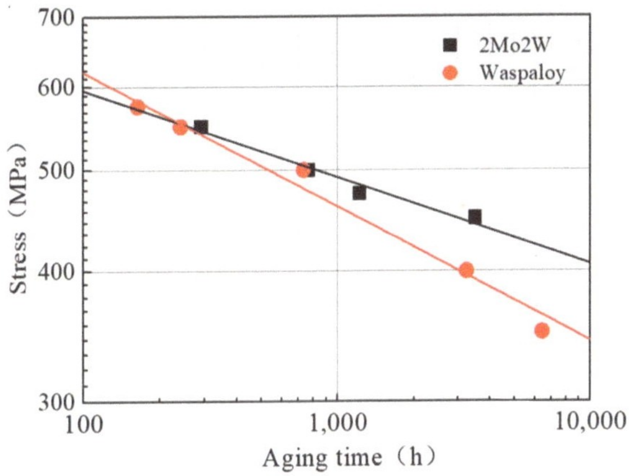

Figure 8. Stress–time creep logarithmic plots for 2Mo2W and Waspaloy.

The strengthening modes of the Ni-based alloy include solid-solution strengthening, γ'-phase precipitation strengthening, and grain boundary precipitation strengthening. To probe the reasons for the creep-fracture strength improvement after replacing Mo with W, the strengths, precipitated phases, and carbide compositions of the two alloys were compared in detail. From Figure 9a,b, one can see the tensile strength, yield strength and hardness of the 2Mo2W alloy before and after aging were higher than those of Waspaloy.

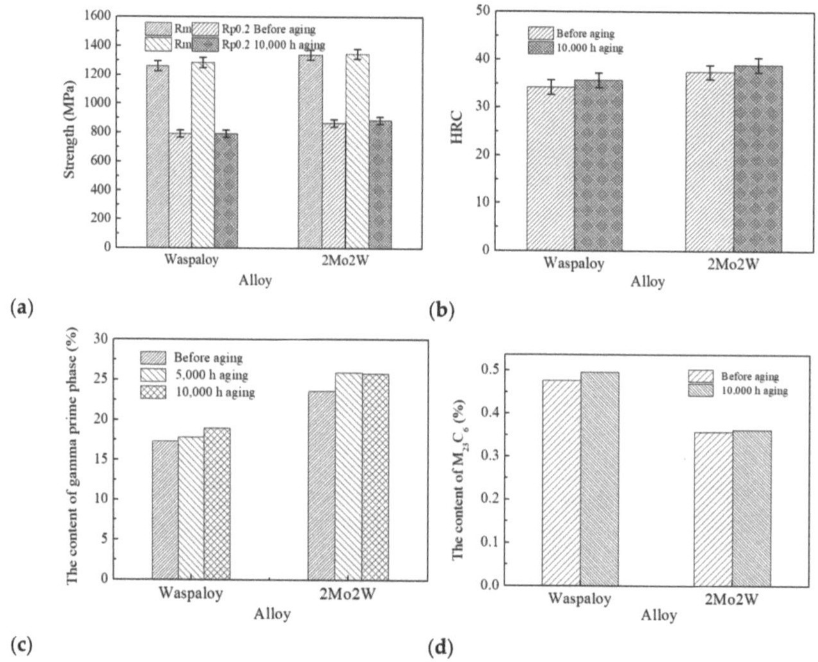

Figure 9. Strengths and precipitate compositions of 2Mo2W alloy and Waspaloy. (**a**) Strength; (**b**) Hardness; (**c**) The content of the γ' phase; (**d**) The content of $M_{23}C_6$ carbide.

The γ' phase contents of the two alloys prior to aging and after 5000 h and 10,000 h are compared in Figure 9c. The γ' phase content in the 2Mo2W alloy was significantly higher than that in the Waspaloy. After 10,000 h of aging, the γ' phase content of the 2Mo2W alloy was about 5% higher than that of Waspaloy. Figure 9d shows that the $M_{23}C_6$ carbide contents of the two alloys were not significantly different.

Mainly due to the large atomic radius of the added W, additional lattice distortion is generated, which increases the resistance to dislocation motion and increases the solid-solution strengthening effect. Meanwhile, more of the γ' phase leads to an increased precipitation-strengthening effect, imparting the new 2Mo2W alloy with greater strength.

The model of critical shear stress was used to estimate the precipitation-strengthening effect of the γ' phase in the Ni-based alloy, which was mainly estimated by the cutting mechanism or the critical shear stress induced by the bypass mechanism when the γ' phase acts by dislocation.

In the cutting mechanism, for small particles, the weakly coupled dislocation (WCD) model is applicable, whereas for large particles, the strongly coupled dislocation (SCD) model is applicable [22,23].

In both models, the critical shear stress is determined as that required for the edge dislocation pair to cut through the γ' phase along the <110> direction in the plane (111). In the WCD model, the critical shear stress can be described as:

$$\Delta\tau = \frac{1}{2}A(\frac{\Gamma}{b})^{\frac{3}{2}}(\frac{bdf}{T})^{\frac{1}{2}} - \frac{1}{2}(\frac{\Gamma}{b})f \tag{1}$$

where A is the shape factor of the particle, 0.72 is the value of the spherical particles, Γ is the antiphase domain bounding energy of the γ' phase in the (111) plane, b is the Bergs vector of the edge dislocation in the matrix, and f and d are the volume fraction and diameter of the γ' phase, respectively. T is the linear tension of the dislocation, which can be calculated as follows:

$$T = 0.5Gb^2 \tag{2}$$

where G is the shear modulus, taken as 68 GPa. In the WCD model, the critical shear stress can be described as:

$$\Delta\tau = \frac{1.72}{2}(\frac{Tf^{\frac{1}{2}}\omega}{bd})(\frac{1.28d\Gamma}{\omega T} - 1)^{\frac{1}{2}} \tag{3}$$

where ω is a constant unit.

In the bypass mechanism (Orowan), the critical shear stress can be described as:

$$\Delta\tau = \frac{Gb}{L} \tag{4}$$

where L is the spacing of the nearest neighbor atom, which can be calculated by:

$$L = \frac{2(1-f)d}{3f} \tag{5}$$

In the calculation, Γ was set at 0.28 J/m^2 and b was set at 0.254 nm. The volume fraction f was taken as 18% for Waspaloy and 25% for 2Mo2W. The calculated data were substituted into the above three models to calculate the corresponding critical shear stresses, as shown in Figure 10.

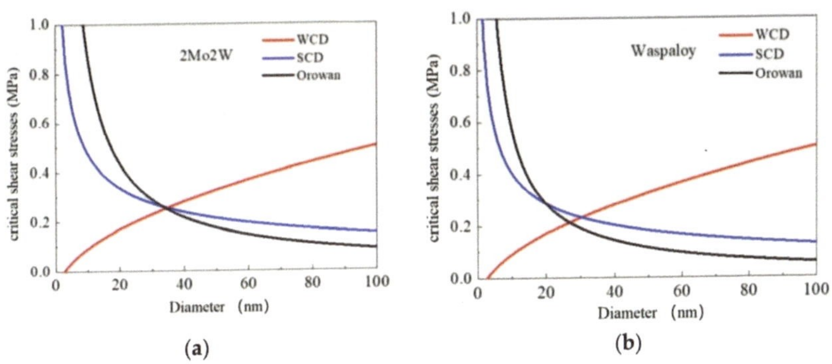

(**a**) (**b**)

Figure 10. Fitting curves of theoretical critical shear stress versus diameter of the γ' phase for 2Mo2W (**a**) and Waspaloy (**b**) at 700 °C.

Given the diameter d of the γ' phase particles, the critical shear stresses corresponding to the three models can be calculated. According to the principle of energy minimization, the dislocation preferentially moves in the direction of least resistance. Therefore, for a given value of d, the actual critical shear stress is the minimum in the three models.

As can be seen from Figure 10, both alloys are dominated by the WCD model when the particles are at their initial size. Figure 10a shows that when 2Mo2W particles grow to about 35 nm, the dislocation bypass model plays a dominant role, and the critical shear stress at this stage is 0.355 MPa. Figure 10b shows that Waspaloy is similar to 2Mo2W, with only two mechanisms of strong coupling, the dislocation model and the Orowan dislocation bypass model, and the critical shear stress is 0.208 MPa.

Compared with the above results, for the same particle radius of the γ' phase, the critical shear stress of 2Mo2W is larger and the strengthening effect is better. This explains why the increase of the γ' phase content is the main reason for the high creep-fracture strength of 2Mo2W.

The stability of the γ' phase can also affect the creep-fracture performance of Ni-based alloys [24], and the γ' phase coarsening rate and the mismatch degree of the γ' and γ phases (matrix) of the 2Mo2W alloy and Waspaloy were compared.

Small lattice mismatch is beneficial to the creep-fracture properties of the alloy, which is expressed by δ, defined as follows:

$$\delta = 2\frac{\alpha_{\gamma'} - \alpha_{\gamma}}{\alpha_{\gamma'} + \alpha_{\gamma}} \tag{6}$$

Figure 11 shows the results of dynamics software calculations of the mismatch degrees of the γ and γ' phases in alloys with different W and Mo contents. It can be seen that an alloy with 4% W and no Mo showed the greatest degree of mismatch, followed by an alloy with 4.5% Mo and no W. The least mismatch was seen for the new alloy with 2% W and 2% Mo. This corroborates that the compound addition of Mo and W is conducive to reducing the mismatch of the γ and γ' phases, stabilizing the γ' phase, consistent with the finding that the coarsening rate of the γ' phase of 2Mo2W alloy is lower than that of Waspaloy.

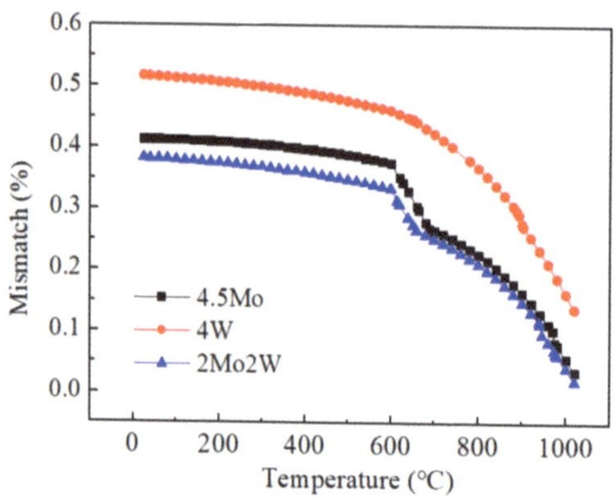

Figure 11. Mismatch of the γ and γ' phases for alloys with different W and Mo contents.

4. Conclusions

A new alloy, designated as 2Mo2W, in which Mo is partially replaced by W, is proposed. During long-term heating of this alloy at 700 °C, the main precipitates that separate have been identified as $M_{23}C_6$ and MC. The creep-fracture strength of 2Mo2W is higher than that of Waspaloy after 10,000 h, which can be mainly attributed to the compound addition of W and Mo, more of the precipitation-strengthening γ' phase, and reduced mismatch of the γ and γ' phases. This renders the γ' phase more stable and lowers the coarsening rate during long-term service at 700 °C. Since the atomic radius of W is larger than that of Mo, its addition reduces the diffusion rates of other alloying elements, increases the stability of the microstructure, and thus improves the creep-fracture strength of the alloy.

Author Contributions: Conceptualization, H.Y. and J.D.; methodology, H.Y. and J.Z.; software, validation, formal analysis, resources, and investigation data curation, H.Y.; writing—original draft preparation, writing—review and editing, H.Y.; supervision, project administration, and funding acquisition, G.Y. and Z.G. All authors have read and agreed to the published version of the manuscript.

Funding: This research was funded by the Inner Mongolia Natural Science Foundation (No. 2022MS0 5039).

Data Availability Statement: Not applicable.

Conflicts of Interest: The authors declare no conflict of interest.

References

1. Liu, Z.; Chen, Z.; He, X.; Bao, H. Systematical Innovation of Heat Resistant Materials Used for 630~700 °C Advanced Ultra-Supercritical (A-USC) Fossil Fired Boilers. *Acta Metall. Sin.* **2020**, *56*, 539–548. (In Chinese)
2. Du, J.; Lv, X.; Dong, J.; Sun, W.; Bi, Z.; Zhao, G.; Deng, Q.; Cui, C.; Ma, H.; Zhang, B. Research progress of wrought superalloys in China. *Acta Metall. Sin.* **2019**, *55*, 1115. (In Chinese)
3. Abe, F.; Kutsumi, H.; Haruyama, H.; Okubo, H. Improvement of oxidation resistance of 9 mass% chromium steel for advanced-ultra supercritical power plant boilers by pre-oxidation treatmentOriginal Research Article. *Corros. Sci.* **2017**, *114*, 1–9. [CrossRef]
4. Wang, L.; Yang, G.; Liu, Z.; Wang, L.; Ma, L.; Yang, Z. Effects of Long-term Aging on Microstructure and Mechanical Properties of a Nickel-base Alloy. *Rare Met. Mater. Eng.* **2018**, *47*, 961–967.
5. Peng, T.; Yang, B.; Yang, G.; Wang, L.; Gong, Z. Stress rupture properties and deformation mechanisms of Nimonic 105 alloy at intermediate temperature. *Mater. Sci. Eng. A* **2020**, *777*, 139085. [CrossRef]
6. Kelekanjeri, V.S.K.G.; Moss, L.K.; Gerhardt, R.A.; Ilavsky, J. Quantification of the coarsening kinetics of γ' precipitates in Waspaloy microstructures with different prior homogenizing treatments. *Acta Mater.* **2009**, *57*, 4658–4670. [CrossRef]
7. Chamanfar, A.; Jahazi, M.; Gholipour, J.; Wanjara, P.; Yuea, S. Evolution of flow stress and microstructure during isothermal compression of Waspaloy. *Mater. Sci. Eng. A* **2014**, *615*, 497–510. [CrossRef]
8. Razumovskiy, V.I.; Lozovoi, A.Y.; Razumovskii, I.M. First-principles-aided design of a new ni-base superalloy: Influence of transition metal alloying elements on grain boundary and bulk cohesion. *Acta Mater.* **2015**, *82*, 369–377. [CrossRef]
9. Bao, H.-S.; Yang, G.; Chen, Z.-Z.; Liu, Z.-D. Effects of long-term aging on microstructure and properties of a tungsten bearing heat-resistant alloy. *J. Iron Steel Res. Int.* **2020**, *27*, 477–487. [CrossRef]
10. Gong, Z.-H.; Ma, Y.-Y.; Bao, H.-S.; Yang, G. Effect of W on formation and properties of precipitates in Ni-based superalloys. *J. Iron Steel Res. Int.* **2021**, *28*, 910–919. [CrossRef]
11. Gong, Z.; Bao, H.; Yang, G. Dynamic Recrystallization and Hot-Working Characteristics of Ni-Based Alloy with Different Tungsten Content. *Metals* **2019**, *9*, 298. [CrossRef]
12. Bao, H.S.; Gong, Z.H.; Chen, Z.Z.; Yang, G. Evolution of precipitates in Ni–Co–Cr–W–Mo superalloys with different tungsten contents. *Rare Met.* **2020**, *39*, 716–724. [CrossRef]
13. Wang, L.; Yang, G.; Lei, T.; Yin, S.-B.; Wang, L. Hot Deformation Behavior of GH738 for A-USC Turbine Blades. *J. Iron Steel Res. Int.* **2015**, *22*, 1043–1048. [CrossRef]
14. Wang, L.; Yang, G.; Lei, T.; Yin, S.-B.; Wang, L. Mechanical Properties and Microstructure of Waspaloy Superalloy with W Additions. *Chin. J. Rare Met.* **2016**, *40*, 117–124.
15. Liu, T.F.; Peng, S.W.; Lin, Y.L.; Wu, C.C. Orientation relationships among M23C6, M6C, and austenite in an Fe-Mn-Al-Mo-C alloy. *Metall. Trans. A* **1990**, *21*, 567–574. [CrossRef]
16. Lvov, G.; Levit, V.I.; Kaufman, M.J. Mechanism of Primary MC Carbide Decomposition in Ni-Base Superalloys. *Metall. Mater. Trans. A* **2004**, *35*, 1669–1679. [CrossRef]
17. Qin, X.Z.; Guo, J.T.; Yuan, C.; Chen, C.L.; Ye, H.Q. Effects of long-term thermal exposure on the microstructure and properties of a cast Ni-base superalloy. *Metall. Mater. Trans. A* **2007**, *38*, 3014–3022. [CrossRef]
18. Wang, J.; Zhou, L.; Qin, X.; Sheng, L.; Hou, J.; Guo, J. Primary MC decomposition and its effects on the rupture behaviors in hot corrosion resistant Ni-based superalloy K444. *Mater. Sci. Eng. A* **2012**, *553*, 14–21. [CrossRef]
19. Wu, X.; Li, Y.; Huang, M.; Liu, W.; Hou, Z. Precipitation kinetics of ordered γ'phase and microstructureevolution in a Ni-Al alloy. *Mater. Chem. Phys.* **2016**, *182*, 125–132. [CrossRef]
20. Turchi, P.E.A.; Kaufman, L.; Liu, Z.K. Modeling of Ni–Cr–Mo based alloys: Part II—Kinetics. *Comput. Coupling Phase Diagr. Thermochem.* **2007**, *31*, 237–248. [CrossRef]
21. Brown, L.; Ham, R. *Strengthening Methods in Crystals*; Applied Science Publication Ltd.: London, UK, 1971.
22. Reppich, B.; Schepp, P.; Wehner, G. Some new aspects concerning particle hardening mechanisms in γ' precipitating nickel-base alloys—II. Experiments. *Acta Metall.* **1982**, *30*, 95–104. [CrossRef]
23. Masoumi, F.; Jahazi, M.; Shahriari, D.; Cormier, J. Coarsening and dissolution of γ' precipitates during solution treatment of AD730™ Ni-based superalloy: Mechanisms and kinetics models. *J. Alloy. Compd.* **2016**, *658*, 981–995. [CrossRef]
24. Ricks, R.A.; Poter, A.J.; Ecob, R.C. The growth of gamma prime precipitates in nickel-base superalloys. *Acta Metall.* **1983**, *31*, 43–53. [CrossRef]

metals MDPI

Article

Effect of Laser Power on the Recrystallization Temperature of an Additively Manufactured IN718

Deuk Hyun Son [1,2], In Soo Kim [2], Baig-Gyu Choi [2], Jeonghyeon Do [2], Yoon Suk Choi [1,*] and Joong Eun Jung [2,*]

[1] School of Materials Science and Engineering, Pusan National University, Busan 46241, Republic of Korea; dhson0929@naver.com

[2] High Temperature Materials Research Group, Korea Institute of Materials Science, Changwon 51508, Republic of Korea; kis@kims.re.kr (I.S.K.); choibg@kims.re.kr (B.-G.C.); jhdo84@kims.re.kr (J.D.)

[*] Correspondence: choiys@pusan.ac.kr (Y.S.C.); diskbios@kims.re.kr (J.E.J.); Tel.: +82-51-510-2382 (Y.S.C.); +82-55-280-3491 (J.E.J.)

Abstract: Over the past few decades, there has been much research on additive manufacturing in both the academic and the industrial spheres to overcome the limitations of conventional manufacturing methods, thereby enabling the production of complex designs for improved performance. To achieve this purpose, it is crucial to meticulously set suitable laser parameters within the context of microstructural characteristics, including type and fraction of defects, texture development, residual stress, and grain size, etc. In the present study, we focused on recrystallization behavior, a type of relaxation process for accumulated thermal stress during the L-PBF process, as a function of laser power applied on the L-PBF process. The laser power has significant effects on the amount of recrystallized grain, directly related to the recrystallization temperature. Within the range of laser power used in this study, a downward trend was observed in the recrystallization temperature as the laser power increased from 370 W to 390 W. This trend suggests that higher laser power leads to a faster cooling rate, influenced by the volume of melt pool as well as the amount of heat dissipation from the melt pool, resulting in higher thermal stress during the process.

Keywords: recrystallization temperature; driving force for recrystallization during laser powder bed fusion; laser powder bed fusion; IN718; thermal stress

1. Introduction

Additive manufacturing of metallic materials has recently drawn much interest in both research and industrial fields due to its potential applications in various high-value-added components, such as gas turbine components, cranial prostheses, dental implants, and more [1–3]. In particular, fabrication of gas turbine components through additive manufacturing can reduce time and cost compared to conventional methods, while enhancing performance of the components. Therefore, gas turbine original equipment manufacturers (OEMs) are actively conducting research on additive manufacturing of Ni-based superalloys, Ti alloys, and Co alloys with various manufacturing techniques, such as laser powder bed fusion (L-PBF), direct energy deposition (DED), and electron beam melting (EBM) [2,4–7]. Among the precipitation-strengthened alloys, the IN718 alloy could provide excellent weldability due to its low Al and Ti content. In addition, it is widely versatile due to its superior heat resistance and oxidation resistance, making it the subject of extensive research in additive manufacturing of Ni-based superalloys.

Energy density is widely used in different forms by researchers, but it is mostly employed in the design of manufacturing processes for fabrication of structures with optimal density, predicting the types of defects that arise during additive manufacturing [8–10]. However, a higher density of a material does not necessarily guarantee superior mechanical properties, such as tensile strength, creep life, and ductility. The mechanical properties of an additively manufactured material are influenced not only by the fraction of the defect but also by factors such as texture, dislocation density, grain size, and so on. One of the crucial phenomena of microstructural evolution is recrystallization [11,12]. It is the thermal stress or dislocation formed during the process that acts as the driving force for recrystallization of the material [13,14]. As a result, numerous strain-free grains are formed which have a considerable effect on the strength and ductility of materials [15–17]. Recently, the effects of volumetric energy density (VED) on the density, grain morphology, and residual stress variations of L-PBF IN718 across a wide range of 29 to 177 J/mm^3 have been reported with the analysis of lattice distortion and residual stress by using EBSD and XRD techniques [18,19]. With increasing VED, lattice distortion was found to increase, leading to a general tendency of increased KAM (kernel average misorientation) values in EBSD analyses. Furthermore, researchers interpreted the occurrence of residual stress in the as-built state as a result of thermal gradients during the additive manufacturing process. A notable finding in these results is that as VED increased in the range of 38 to 90 J/mm^3, the residual stress also increased. The edges of the specimens exhibited residual stresses approximately two to five times higher than those in the central regions, indicating that a higher cooling rate promoted higher residual stress. However, the effects of laser parameters on the driving force for recrystallization as well as the recrystallization temperature of additively manufactured IN718 has not been clarified to date. Previous reports did not analyze the changes in recrystallization temperature according to VED or laser parameters. Given that international standards, such as ASTM and AMS, provide guidelines for stress-relief heat treatment conditions for L-PBF IN718, it is essential to investigate the variations in recrystallization temperature with respect to laser parameters. Therefore, this study attempts to investigate the effect of laser power on the driving force for the recrystallization, which in turn affects the recrystallization temperature of additively manufactured IN718.

2. Materials and Methods

IN718 Powder and Laser Powder Bed Fusion

The spherical IN718 powders with sizes ranging from 15 to 53 μm, manufactured by AP&C with the electrode inert gas atomization (EIGA) process, were used. The chemical composition of the IN718 powder is shown in Table 1.

Table 1. Chemical composition of IN718 powder.

Wt. %	C	Cr	Co	Mo	Nb	Ti	Al	Ni	Si	Mn	P	S	Fe
	0.05	18.5	<0.1	2.95	5.11	0.97	0.51	52.45	0.07	0.06	0.004	<0.001	Bal.

The Concept Laser M2 machine (GE Additive, Cincinnati, OH, USA), capable of utilizing dual laser beams, was employed for the L-PBF process to fabricate IN718 specimens. The detailed laser parameters, as shown in Table 2, involved varying the laser power between 370 W, 380 W, and 390 W, while keeping the laser scan speed, hatch distance, and layer thickness constant at 600 mm/s, 0.1 mm, and 25 μm, respectively. To minimize anisotropy as well as thermal stress of the specimen, the laser scan direction was rotated by 67 degrees for each layer. A support structure with a height of 5 mm was set up, and a cylindrical specimen with a diameter of 13 mm and a length of 80 mm was built on the support structure. As-built specimens were wire-cut from the base plate, each specimen was divided into eight equal parts with 10 mm intervals, as shown in Figure 1, denoted as bottom (No. 1,2), middle (No. 4,5), and top (No. 7,8), to observe the porosity fraction and initial microstructure according to the sample position. The porosity fraction was quantified using image analysis (Image-Pro Plus, Media Cybernetics, Rockville, MD, USA) after grinding with sandpaper, polishing with 1 μm diamond paste, and capturing optical microscope images at 10× magnification without etching. The amount of residual stress in the as-built specimens was examined by the XRD -sin$^2\psi$ method using XStress-3000 G3 [20,21]. This method could evaluate the residual stress from the change of lattice spacing with different tilting angle (ψ). The residual stress was calculated based on the following equation,

$$\frac{d_{\Phi\Psi} - d_0}{d_0} = \left(\frac{1+\nu}{E}\right)\cdot\sigma_\Phi\cdot sin^2(\Psi) - \frac{\nu}{E}\cdot(\sigma_1 + \sigma_2) \quad (1)$$

where $d_{\Phi\psi}$, d_0, E, and ν stand for the lattice spacing of the (hkl) plane, lattice spacing of stress-free state, the Young's modulus, and Poisson's ratio, respectively. Equation (1) can be rewritten as follows by assuming that $\sigma_1 = \sigma_2$, and σ_Φ is independent of orientation.

$$d_{\Phi\Psi} = \left[\left(\frac{1+\nu}{E}\right)\cdot sin^2(\Psi) - \left(\frac{2\nu}{E}\right)\right]\cdot\sigma\cdot d_0 + d_0 \quad (2)$$

Therefore, it is possible to establish a linear explanation between $d_{\Phi\psi}$ (= $\lambda/2\sin\theta$) and $\left[\left(\frac{1+\nu}{E}\right)\cdot sin^2(\Psi) - \left(\frac{2\nu}{E}\right)\right]$, which in turn allows us to obtain values of in-plane stress (σ) and d_0. In this study, Cr-Kα radiation was used, and eight different tilting angles ranging from −45 degrees to +45 degrees were employed around the $(220)_\gamma$ peak. The stress-relief heat treatment was conducted at 1065 °C, 1080 °C, and 1095 °C for 90 min to avoid precipitation of δ-phase (Ni_3Nb, $D0_a$), with a reference standard heat treatment following ASTM F3055 [22]. The aging heat treatment was performed according to AMS 5596, with a hold at 718 °C for 8 h [23]. It was then furnace-cooled down to 621 °C and held for an additional 8 h, followed by air cooling. The recrystallization after heat treatment was examined by performing electron backscatter diffraction (EBSD) (Oxford instruments, Abingdon, UK) with an acceleration voltage of 15 kV, and a step size of 4 μm. The observation was carried out on an area of 2560 × 1900 μm^2 of the plane normal to the building direction, as shown in Figure 1. The acquired data were post-processed using TSL-OIM software (version 7.3.1, EDAX, Mahwah, NJ, USA) with a filtering criterion applied to utilize only values with confidence index greater than 0.1 and grain size exceeding 3 μm.

Table 2. Laser powder bed fusion parameters for IN718 used in the present study.

Laser Power	Spot Size	Scan Speed	Hatch Distance	Layer Thickness	Atmosphere
370 W 380 W 390 W	180 μm	600 mm/s	100 μm	25 μm	Ar purging

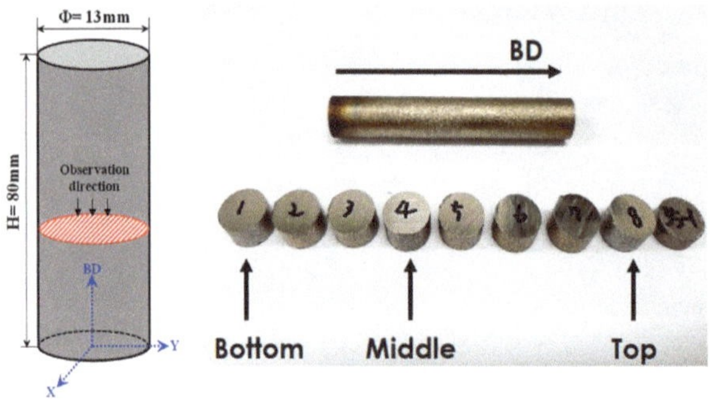

Figure 1. Sample location and observation direction for microstructure analysis.

Based on the grain orientation spread (GOS) value, grains with less than 1 degree deviation from the mean orientation were considered to be recrystallized grain. Finally, the fraction of the recrystallized grains was deduced by calculating the area fraction of grains with the GOS value under 1 degree. The dislocation cell structure of the as-built specimen depending on laser power was observed using a transmission electron microscope (TEM, JEM-2100F, JEOL, Tokyo, Japan). The TEM sample was prepared with a 3 mm disc of 50 μm in thickness, followed by jet-polishing. The observation was performed after tilting the sample, aligning the incident beam parallel to the $<001>_{\gamma}$ direction.

3. Results

3.1. Microstructure of As-Built Specimens

The porosity was measured on the plane perpendicular to the building direction (XY plane) of bottom, middle, and top region of each specimen. For the 370 W condition, the specimens were measured to have 0.621%, 0.787%, 1.263% of porosity at the bottom, middle, and top regions, respectively. Under the 380 W condition, the measurements were 2.457%, 3.597%, 4.955% at the bottom, middle, and top regions, respectively. Additionally, for the 390 W condition, the measurements were 3.201%, 2.581%, 2.914% at the bottom, middle, and top regions, respectively. The energy densities under these three conditions were higher than the optimal energy density provided by the powder supplier, indicating an increase in the key-hole fraction compared to defect fraction under the optimum laser parameter set [24]. Figure 2 presents the melt pool size and shape of the as-built specimens, observed at the plane perpendicular to laser scan direction. The average width and depth of the melt pool, measured at more than 10 different points, are denoted in the top-left corner of each optical micrograph. Generally, it is reported that the width and depth of the melt pool tend to increase simultaneously with increasing laser power, resulting in a larger volume. However, in this experiment, it was found that the width remained nearly constant or decreased slightly, whereas the depth of melt pool continuously increased as the laser power increased. As a result, the shape of melt pool changes from a shallow shape to a deep shape while maintaining the width.

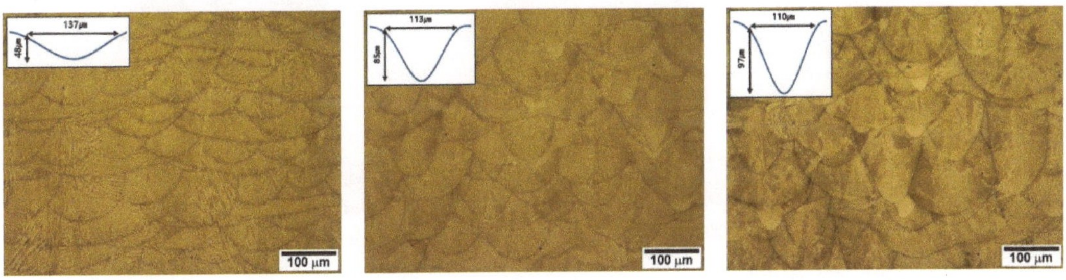

(a) (b) (c)

Figure 2. Melt pool size and shape observed at the plane normal to the scan direction: (**a**) 370 W; (**b**) 380 W; (**c**) 390 W.

Figure 3 illustrates the EBSD results obtained from the bottom region of the as-built specimens. One could notice from the inverse pole figure (IPF) map that multiple grains exhibit a regular alignment along the laser scan direction, which was rotated by 67 degrees every single layer. The positions of the observed (001) poles in the pole figure indicate that (001) poles are located either parallel or perpendicular to the laser scan direction. Moreover, a weak cube texture is discernible in the as-built specimens, and the maximum intensity of the poles diminishes as the laser power increases. However, this characteristic can vary depending on the surface location examined in EBSD imaging, whether it is at the intermediate depth of the layer or at the boundary between layers. Therefore, it is difficult to find the exact relationship between laser power and texture development in the present results. Nevertheless, considering the relationship between the laser scan direction and the position of the (001) poles, it is inferred that the overall as-built specimen exhibits a {001} fiber texture.

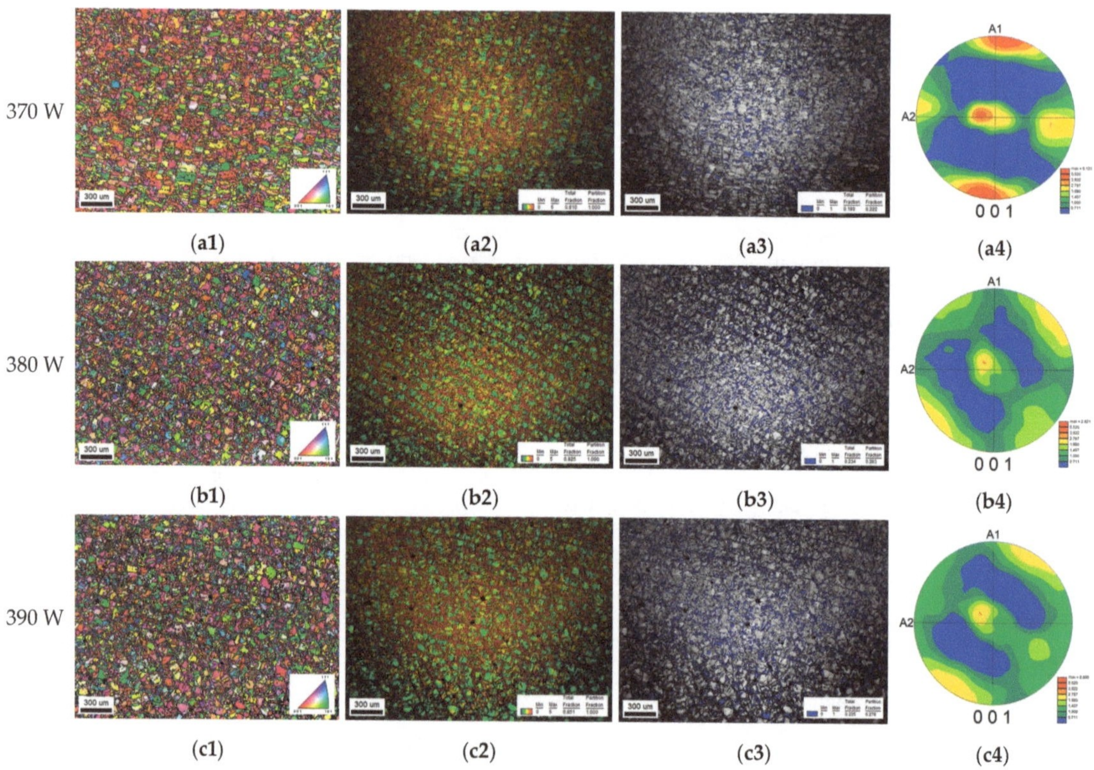

Figure 3. Electron backscattered diffraction images at the XY plane of as-built specimens: (**a1**,**b1**,**c1**) inverse pole figure map; (**a2**,**b2**,**c2**) kernel average misorientation map; (**a3**,**b3**,**c3**) image quality map with highlighting GOS under 1 degree; and (**a4**,**b4**,**c4**) pole figure map.

The KAM is a measure of the difference in orientation between a certain point and its surroundings. By maintaining a constant step size, KAM can be used as an indication of the relative deformation received during the additive manufacturing process. In the KAM map, alternating green and red bands represent low and high KAM values, respectively. This alternating pattern is also associated with the width of the melt pool and the hatch distance. Specifically, during the solidification of the melt pool, the surrounding region experiences tensile stresses due to solidification shrinkage of the melt pool. Consequently, higher deformation occurs in the boundary region of the melt pool compared to its central part, resulting in higher KAM values at the edges of the melt pool. However, the average KAM values measured in the observed area showed no significant trend with respect to laser power, with values of 3.248, 3.314, and 3.141 for 370, 380, and 390 W, respectively. Therefore, there are limitations in determining the driving force for recrystallization with increasing laser power based on the EBSD test condition performed in this study.

The thermal stresses generated during the L-PBF process can persist elastically even after plastic deformation occurred during rapid cooling [25–29]. This residual stress was measured in the bottom region of specimens fabricated under conditions of 370 W, 380 W, and 390 W, with the corresponding results depicted in Figure 4. It was observed that all measured specimens exhibited tensile residual stress, and as the laser power increased from 370 W to 390 W, with a 20 W increment, the residual stresses escalated from 195.6 MPa to 433.3 MPa. This result indicates the occurrence of higher thermal stresses with increasing laser power.

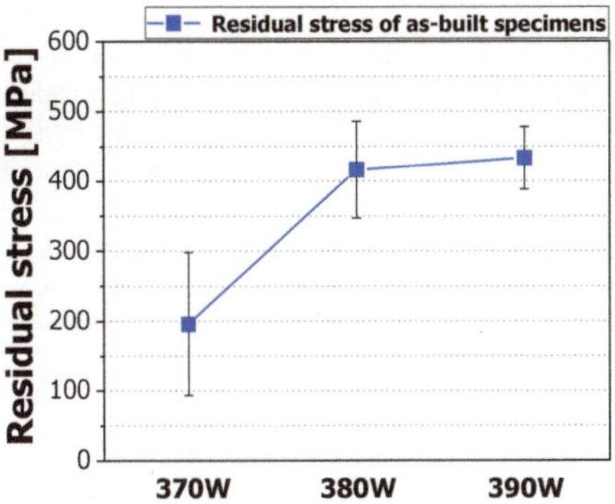

Figure 4. Residual stress of the as-built specimens measured from the bottom region.

3.2. Microstructure of Heat-Treated Specimens

The bottom region of the as-built specimens was subjected to stress-relief heat treatment in an argon atmosphere at temperatures of 1065 °C, 1080 °C, and 1095 °C, and the resulting microstructural changes were observed. Figure 5 presents the EBSD microstructure observations for a laser power of 370 W with increasing stress-relief heat treatment temperatures. Even after heat treatment, the laser scan path remains visible in both the IPF map and the KAM map. In addition, there was no significant change in the average KAM value within the measured area as the heat treatment temperature increased, with values of 2.887, 3.037, and 3.148. Furthermore, the recrystallized grain fraction based on the GOS value was 22.9%, 18.0%, and 16.9%, respectively. Assuming that the recrystallization temperature is where more than 50% of the total area undergoes recrystallization, the recrystallization temperature under the laser power condition of 370 W was found to be above 1095 °C.

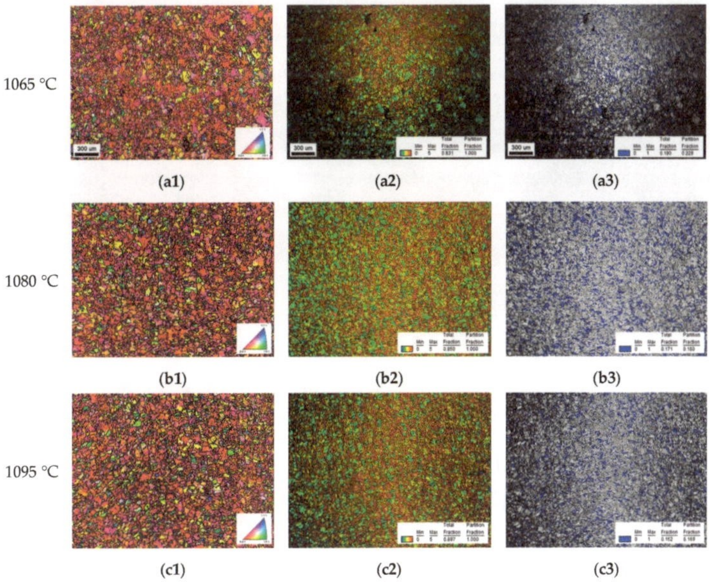

Figure 5. EBSD images for 370 W specimens with different stress-relief heat treatment temperatures: (**a1**,**b1**,**c1**) inverse pole figure map; (**a2**,**b2**,**c2**) kernel average misorientation map; (**a3**,**b3**,**c3**) image quality map highlighting GOS under 1 degree.

Similarly, specimens manufactured using a laser power of 380 W underwent heat treatment at the same three temperatures, and the resulting EBSD microstructure observations are displayed in Figure 6. Under the 380 W condition, the laser scan path remained clearly visible across all heat treatment conditions. The average KAM values for the entire map were found to be 3.103, 3.104, and 2.796, respectively, as the temperature increased. Notably, after heat treatment at 1080 °C and 1095 °C, the occurrence of local recrystallization became evident, as indicated by the yellow arrows in Figure 6a1–b3. Additionally, the recrystallized grain fractions based on the GOS values measured at 1065 °C, 1080 °C, and 1095 °C were 41.4%, 33.3%, and 47.1%, respectively. It is inferred that the recrystallization temperature for specimens under the laser power condition of 380 W approaches 1095 °C.

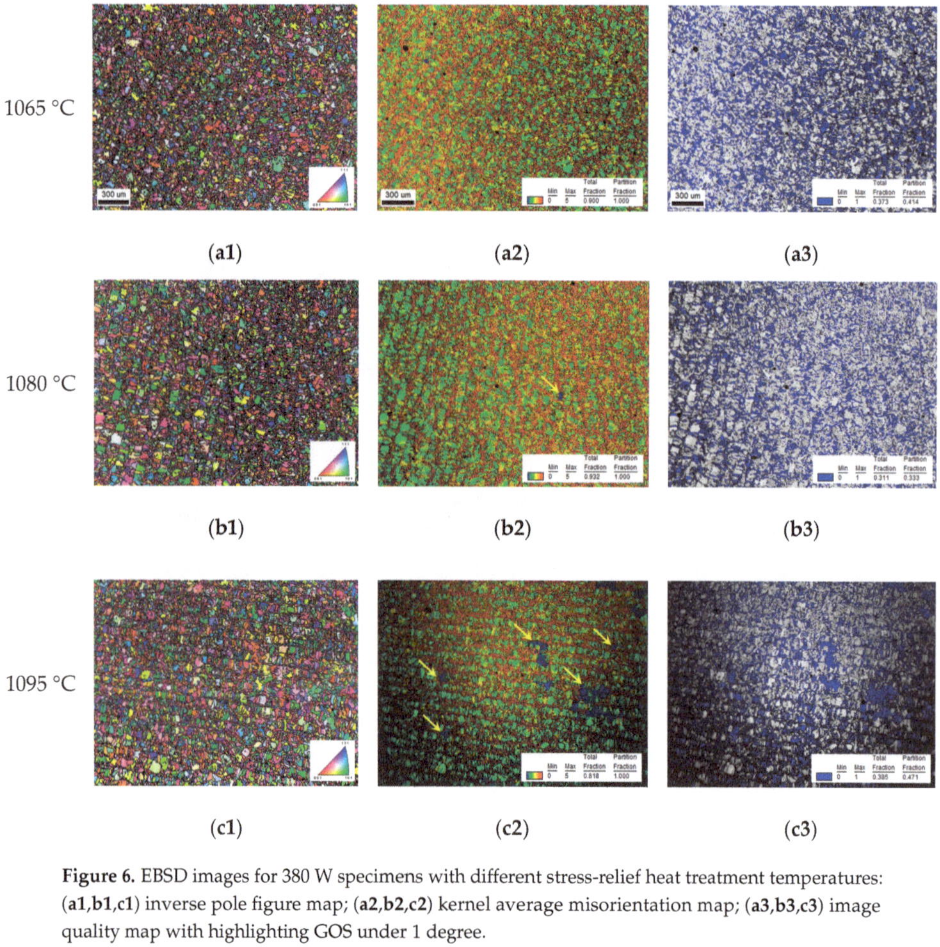

Figure 6. EBSD images for 380 W specimens with different stress-relief heat treatment temperatures: (**a1**,**b1**,**c1**) inverse pole figure map; (**a2**,**b2**,**c2**) kernel average misorientation map; (**a3**,**b3**,**c3**) image quality map with highlighting GOS under 1 degree.

On the other hand, the specimens produced with a laser power of 390 W maintained a lower recrystallized grain fraction of 10.3% after heat treatment at 1065 °C, as shown in Figure 7. In contrast, after heat treatment at 1080 °C and 1095 °C, a significant increase in the fraction of recrystallized grains can be found with values of 85.4% and 94.4%, respectively. As recrystallization progressed, the average KAM value within the measured area exhibited a significant decrease, reaching 3.332, 0.962, and 0.351 for 1065 °C, 1080 °C, and 1095 °C, respectively. Hence, it was confirmed that under the laser power condition of 390 W, the recrystallization temperature of the specimen exists between 1065 °C and 1080 °C.

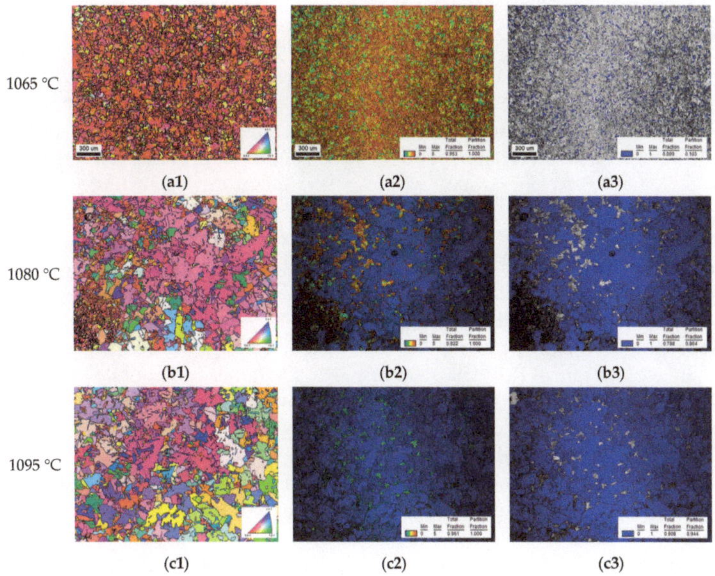

Figure 7. EBSD images for 390 W specimens with different stress-relief heat treatment temperatures: (**a1**,**b1**,**c1**) inverse pole figure map; (**a2**,**b2**,**c2**) kernel average misorientation map; (**a3**,**b3**,**c3**) image quality map with highlighting GOS under 1 degree.

Figure 8 demonstrates the fraction of recrystallized grain as a function of laser power and heat treatment temperature. By considering the recrystallization condition as when more than 50% of the measured area is converted to recrystallized grains, recrystallization occurred only at 1080 °C and 1095 °C after heat treatment under 390 W. Thus, it can be inferred that the recrystallization temperature is above 1095 °C for a laser power of 370 W, approximately 1095 °C for a laser power of 380 W, and between 1065 °C and 1080 °C for a laser power of 390 W.

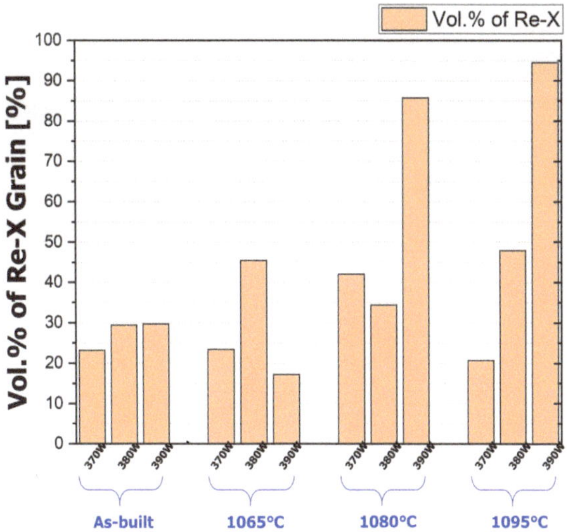

Figure 8. Volume fraction of recrystallized grain as a function of laser power and stress-relief heat treatment temperatures.

4. Discussion

4.1. Driving Force for Recrystallization Accumulated during Laser Powder Bed Fusion

The recrystallization observed in the additively manufactured materials can be attributed to various driving forces [28,30,31]. These forces encompass tensile deformation caused by solidification shrinkage around the melt pool, as well as accumulated tensile and compressive deformations arising from repetitive heating and cooling cycles. The resulting thermal stresses should be relaxed through various relaxation processes, including plastic deformation, solidification cracking, and liquation cracking. However, some stresses within the elastic limit at the particular temperature of a given material may persist as residual stress. Therefore, to comprehend the driving force for recrystallization in relation to the laser power, an evaluation of the factors influencing thermal stress is imperative.

It is widely recognized that the thermal stresses arising during the L-PBF process, under the identical material, equipment, and atmosphere conditions, are considerably influenced by the cooling rate of the melt pool. In essence, a higher cooling rate promotes thermal stress, whereas a relatively lower cooling rate results in reduced thermal stress.

The cooling rate of the melt pool is dependent on the volume of the melt pool [32,33]. Under a constant temperature gradient, an increase in the melt pool's volume prolongs the required solidification time, consequently reducing the cooling rate. Previous studies have indicated that with increasing laser power, both the width and depth of the melt pool tend to increase, leading to a larger volume of the melt pool and, subsequently, a diminished cooling rate. The cooling rate is also influenced by the heat transfer from the melt pool to its surroundings [34,35]. A more pronounced heat transfer from the melt pool can provide a higher cooling rate and rapid solidification. This factor is contingent upon various factors, including the heat capacity of the melt pool, heat transfer mechanisms, heat transfer coefficients, and the surface area per unit volume of the melt pool. Furthermore, the cooling rate is affected by additional factors such as the geometry of the build, support structure, the scan strategy, and the characteristics of the chamber [36–39].

4.2. Driving Force for Recrystallization with Increasing Laser Power

The experimental results confirm that the recrystallization temperature exhibits a downward trend as the laser power increases, indicating an increase in the driving force for recrystallization. It can be expected that the cooling rate of the melt pool increases with increasing laser power. The changes in the dimensions of the melt pool with respect to laser power, as shown in Figure 2, clearly demonstrate that the volume of the melt pool expands with higher laser power, while the width remains constant or slightly decreases, and only the depth experiences an increase. The amount of heat transfer, therefore, tends to increase with a deeper melt pool shape compared to a shallower one. Consequently, it can be deduced that a larger amount of heat dissipation occurs due to the higher surface area per unit volume of the melt pool, resulting in a faster cooling rate with increasing laser power.

To verify this, TEM observation of the as-built specimens at different laser powers were carried out and are presented in Figure 9. By setting the zone axis to $<001>_{\gamma}$, the average cell size was measured, resulting in 707 nm for 370 W, 648 nm for 380 W, and 518 nm for 390 W. These measurement correspond to the increasing trend in heat dissipation of the melt pool with higher laser power, suggesting a faster cooling rate. Additionally, the presence of the dislocation cell structure formed at the cell boundaries was observed. This dislocation cell structure is known to form as a result of repetitive heating and cooling, which leads to tensile and compressive deformation even after solidification. Notably, it can be clearly observed that the dislocation density formed at the cell boundaries increases with increasing laser power.

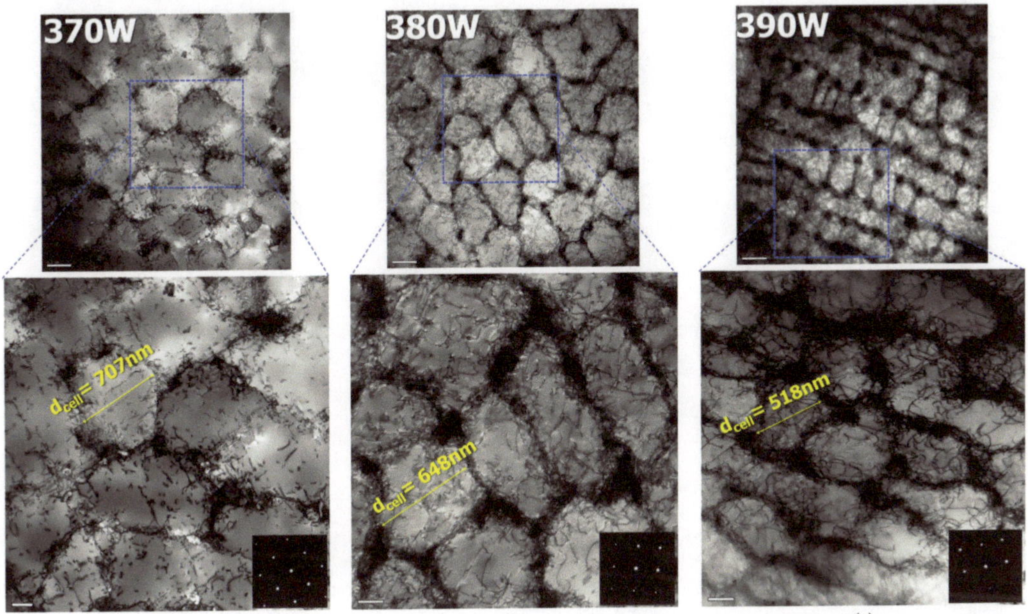

(a) (b) (c)

Figure 9. Dislocation cell structure of the as-built specimens with the zone axis of <001>$_\gamma$: (**a**) 370 W; (**b**) 380 W; (**c**) 390 W.

5. Conclusions

In this study, we investigated the variation in the recrystallization temperature of L-PBF IN718 specimens as a function of laser power. By applying three stress-relief heat treatment temperatures, we were able to clearly distinguish the temperature ranges in which recrystallization occurred under different laser powers. Through various microstructure analyses, we discovered evidence that the reason for the increasing driving force of recrystallization with increased laser power.

1. The observation plane of the as-built specimens exhibited a weak cube texture, with (001) planes aligned perpendicular or parallel to the laser scan path. The average KAM values, under the EBSD conditions used in this study, showed minimal differences based on laser power. However, the residual stresses increased by more than two times as the laser power increased from 370 W to 390 W.
2. By defining the recrystallization temperature based on the area of the recrystallized region after stress-relief heat treatment, we found that the recrystallization temperature was above 1095 °C for 370 W, approximately 1095 °C for 380 W, and within the range of 1065–1080 °C for 390 W. This confirmed that the recrystallization temperature decreased with increasing laser power.
3. Thermal stress resulting from rapid cooling during the L-PBF process is one of the major driving forces for recrystallization. On the basis of the recrystallization temperature, it can be inferred through microstructure analysis that the 390 W condition experienced the fastest cooling rate due to pronounced heat dissipation from the melt pool. TEM observations indeed revealed the smallest cell size under the 390 W condition, indicating a faster cooling rate. Additionally, the highest dislocation density, resulting from repetitive heating and cooling even after solidification, was observed at 390 W, suggesting the highest driving force for recrystallization.

Author Contributions: Conceptualization, J.E.J. and Y.S.C.; methodology, D.H.S., J.E.J. and Y.S.C.; software, B.-G.C. and J.D.; validation, D.H.S., I.S.K., B.-G.C., J.D., Y.S.C. and J.E.J.; formal analysis, I.S.K. and J.D.; investigation, D.H.S. and J.E.J.; resources, D.H.S., J.D. and B.-G.C.; data curation, D.H.S., Y.S.C. and J.E.J.; writing—original draft preparation, D.H.S.; writing—review and editing, J.E.J. and Y.S.C.; visualization, D.H.S. and J.E.J.; supervision, Y.S.C. and J.E.J.; project administration, B.-G.C. and J.E.J.; funding acquisition, B.-G.C. and J.E.J. All authors have read and agreed to the published version of the manuscript.

Funding: This research was funded by Fundamental Research Program (PNK8890) of the Korea Institute of Materials Science (KIMS).

Data Availability Statement: No new data were created or analyzed in this study. Data sharing is not applicable to this article.

Conflicts of Interest: The authors declare no conflict of interest.

References

1. DebRoy, T.; Wei, H.L.; Zuback, J.S.; Mukherjee, T.; Elmer, J.W.; Milewski, J.O.; Beese, A.M.; Wilson-Heid, A.D.; De, A.; Zhang, W. Additive manufacturing of metallic components—Process, structure and properties. *Prog. Mater. Sci.* **2018**, *92*, 112–224. [CrossRef]
2. Blakey-Milner, B.; Gradl, P.; Snedden, G.; Brooks, M.; Pitot, J.; Lopez, E.; Leary, M.; Berto, F.; Du Plessis, A. Metal additive manufacturing in aerospace: A review. *Mater. Des.* **2021**, *209*, 110008. [CrossRef]
3. Dai, K.; Shaw, L. Thermal and stress modeling of multi-material laser processing. *Acta Mater.* **2001**, *499*, 4171. [CrossRef]
4. Giuliani, F.; Paulitsch, N.; Cozzi, D.; Görtler, M.; Andracher, L. An assessment on the benefits of additive manufacturing regarding new swirler geometries for gas turbine burners. In Proceedings of the ASME Turbo Expo 2018, Oslo, Norway, 11–15 June 2018.
5. Prabhakar, P.; Sames, W.J.; Dehoff, R.; Babu, S.S. Computational modeling of residual stress formation during the electron beam melting process for Inconel 718. *Addit. Manuf.* **2015**, *7*, 83–91. [CrossRef]
6. Kok, Y.; Tan, X.P.; Wang, P.; Nai, M.L.S.; Loh, N.H.; Liu, E.; Tor, S.B. Anisotropy and heterogeneity of microstructure and mechanical properties in metal additive manufacturing: A critical review. *Mater. Des.* **2018**, *139*, 565–586. [CrossRef]
7. Park, J.M.; Asghari-Rad, P.; Zargaran, A.; Bae, J.W.; Moon, J.; Kwon, H.; Choe, J.; Yang, S.; Yu, J.H.; Kim, H.S. Nano-scale heterogeneity-driven metastability engineering in ferrous medium-entropy alloy induced by additive manufacturing. *Acta Mater.* **2021**, *221*, 117426. [CrossRef]
8. Han, J.; Yang, J.; Yu, H.; Yin, J.; Gao, M.; Wang, Z.; Zeng, X. Microstructure and mechanical property of selective laser melted Ti6Al4V dependence on laser energy density. *Rapid Prototyp. J.* **2017**, *23*, 217–226. [CrossRef]
9. Li, J.; Wei, Z. Process optimization and microstructure characterization of Ti6Al4V manufactured by selective laser melting. *IOP Conf. Ser. Mater. Sci. Eng.* **2017**, *269*, 012026.
10. Pal, S.; Gubeljak, N.; Hudák, R.; Lojen, G.; Rajťúková, V.; Brajlih, T.; Drstvenšek, I. Evolution of the metallurgical properties of Ti-6Al-4V, produced with different laser processing parameters, at constant energy density in selective laser melting. *Results Phys.* **2020**, *17*, 103186. [CrossRef]
11. Sonis, E.; Dépinoy, S.; Giroux, P.; Maskrot, H.; Lemarquis, L.; Hercher, O.; Villaret, F.; Gourgues-Lorenzon, A. Dependency of recrystallization kinetics of the solidification microstructure of 316L stainless steel processed by laser powder bed fusion (LPBF). *Mater. Char.* **2022**, *194*, 112370. [CrossRef]
12. De Terris, T.; Castelnau, O.; Hadjem-Hamouche, Z.; Haddadi, H.; Michel, V.; Peyre, P. Analysis of As-built microstructures and recrystallization phenomena on Inconel 625 alloy obtained via laser powder bed fusion (L-PBF). *Metals* **2021**, *11*, 619. [CrossRef]
13. Pinto, F.C.; Aota, L.S.; Souza Filho, I.R.; Raabe, D.; Sandim, H.R.Z. Recrystallization in non-conventional microstructures of 316L stainless steel produced via laser powder-bed fusion: Effect of particle coarsening kinetics. *J. Mater. Sci.* **2022**, *57*, 9576–9598. [CrossRef]
14. Jiang, R.; Mostafaei, A.; Wu, Z.; Choi, A.; Guan, P.W.; Chmielus, M.; Rollett, A.D. Effect of heat treatment on microstructural evolution and hardness homogeneity in laser powder bed fusion of alloy 718. *Addit. Manuf.* **2020**, *35*, 101282. [CrossRef]
15. Pröbstle, M.; Neumeier, S.; Hopfenmüller, J.; Freund, L.P.; Niendorf, T.; Schwarze, D.; Göken, M. Superior creep strength of a nickel-based superalloy produced by selective laser melting. *Mater. Sci. Eng. A* **2016**, *674*, 299–307. [CrossRef]
16. Gallmeyer, T.G.; Moorthy, S.; Kappes, B.B.; Mills, M.J.; Amin-Ahmadi, B.; Stebner, A.P. Knowledge of process-structure-property relationships to engineer better heat treatments for laser powder bed fusion additive manufactured Inconel 718. *Addit. Manuf.* **2020**, *31*, 100977. [CrossRef]
17. Zhao, Y.; Meng, F.; Liu, C.; Tan, S.; Xiong, W. Impact of homogenization on microstructure-property relationships of Inconel 718 alloy prepared by laser powder bed fusion. *Mater. Sci. Eng. A* **2021**, *826*, 141973. [CrossRef]
18. Doğu, M.N.; Davut, K.; Obeidi, M.A.; Yalçın, M.A.; Gu, H.; Low, T.S.E.; Ginn, J.; Brabazon, D. Recrystallization and grain growth kinetics of IN718 manufactured by laser powder bed fusion. *J. Mater. Res. Technol.* **2022**, *19*, 4242–4257. [CrossRef]
19. Praveen Kumar, V.; Vinoth Jebaraj, A. Microscale investigations on additively manufactured Inconel 718: Influence of volumetric energy density on microstructure, texture evolution, defects control and residual stress. *App. Phys. A* **2023**, *129*, 370. [CrossRef]

20. Macherauch, E. X-ray stress analysis. *Exptl. Mech.* **1966**, *6*, 140–153. [CrossRef]
21. Mirkoohi, E.; Tran, H.C.; Lo, Y.L.; Chang, Y.C.; Lin, H.Y.; Liang, S.Y. Mechanics modeling of residual stress considering effect of preheating in laser powder bed fusion. *J. Manuf. Mater. Process.* **2021**, *5*, 46. [CrossRef]
22. *ASTM F3055-14a*; Standard Specification for Additive Manufacturing Nickel Alloy (UNS NO7718) with Powder Bed Fusion. ASTM International: West Conshohocken, PA, USA, 2014.
23. *AMS 5596E*; Aerospace Material Specification for Inconel 718 (UNS NO7718) Sheets, Strips and Plates. Society for Automotive Engineers: Warrendale, PA, USA, 1984.
24. Giorgetti, A.; Baldi, N.; Palladino, M.; Ceccanti, F.; Arcidiacono, G.; Citti, P. A method to optimize parameters development in L-PBF based on single and multitracks analysis: A case study on Inconel 718 alloy. *Metals* **2023**, *13*, 306. [CrossRef]
25. Bartlett, J.L.; Li, X. An overview of residual stresses in metal powder bed fusion. *Addit. Manuf.* **2019**, *27*, 131–149. [CrossRef]
26. Levkulich, N.C.; Semiatin, S.L.; Gockel, J.E.; Middendorf, J.R.; DeWald, A.T.; Klingbeil, N.W. The effect of process parameters on residual stress evolution and distortion in the laser powder bed fusion of Ti-6Al-4V. *Addit. Manuf.* **2019**, *28*, 475–484. [CrossRef]
27. Chen, C.; Xiao, Z.; Zhang, W.; Wang, Y.; Zhu, H. Effect of laser jump speed on temperature distribution and thermal stress in laser powder bed fusion. *Opt. Laser Technol.* **2021**, *142*, 107275. [CrossRef]
28. Mirkoohi, E.; Tran, H.C.; Lo, Y.L.; Chang, Y.C.; Lin, H.Y.; Liang, S.Y. Analytical modeling of residual stress in laser powder bed fusion considering part's boundary condition. *Crystals* **2020**, *10*, 337. [CrossRef]
29. Gan, M.; Wu, Q.; Long, L. Prediction of residual deformation and stress of laser powder bed fusion manufactured Ti-6Al-4V lattice structures based on inherent strain method. *Mater. Res.* **2023**, *26*, e20220516. [CrossRef]
30. Ulbricht, A.; Altenburg, S.J.; Sprengel, M.; Sommer, K.; Mohr, G.; Fritsch, T.; Mishurova, T.; Serrano-Munoz, I.; Evans, A.; Hofmann, M.; et al. Separation of the formation mechanisms of residual stresses in LPBF 316L. *Metals* **2020**, *10*, 1234. [CrossRef]
31. Antikainen, A.; Reijonen, J.; Lagerbom, J.; Lindroos, M.; Pinomaa, T.; Lindroos, T. Experimental and Calphad methods for evaluating residual stresses and solid-state shrinkage after solidification. *Metals* **2022**, *12*, 1894. [CrossRef]
32. Zhang, Z.; Ali, U.; Mahmoodkhani, Y.; Huang, Y.; Shahabad, S.I.; Kasinathan, A.R.; Toyserkani, E. Experimental and numerical investigation on the effect of layer thickness during laser powder-bed fusion of stainless steel 17-4 PH. *Int. J. Rapid Manuf.* **2020**, *9*, 212–230. [CrossRef]
33. Farshidianfar, M.H.; Khajepour, A.; Gerlich, A.P. Effect of real-time cooling rate on microstructure in laser additive manufacturing. *J. Mater. Proc. Technol.* **2016**, *231*, 468–478. [CrossRef]
34. Chai, R.; Zhang, Y.; Zhong, B.; Zhang, C. Effect of scan speed on grain and microstructural morphology for laser additive manufacturing of 304 stainless steel. *Rev. Adv. Mater. Sci.* **2021**, *60*, 744–760. [CrossRef]
35. Hyer, H.C.; Petrie, C.M. Effect of powder layer thickness on the microstructural development of additively manufactured SS316. *J. Manuf. Proc.* **2022**, *76*, 666–674. [CrossRef]
36. Chua, Z.Y.; Moon, S.K.; Jiao, L.; Ahn, I.H. Geometric influence of the laser-based powder bed fusion process in Ti6Al4V and AlSi10Mg. *Int. Adv. Manuf. Technol.* **2021**, *114*, 3165–3176. [CrossRef]
37. Song, H.; McGaughy, T.; Sadek, A.; Zhang, W. Effect of structural support on microstructure of nickel base superalloy fabricated by laser powder bed fusion additive manufacturing. *Addit. Manuf.* **2019**, *26*, 30–40. [CrossRef]
38. Ekubaru, Y.; Gokcekaya, O.; Nakano, T. Effects of scanning strategy on the microstructure and mechanical properties of Sc-Zr-modified Al-Mg alloy manufactured by laser powder bed fusion. *Crystals* **2022**, *12*, 1348. [CrossRef]
39. Sun, Q.; Guo, K.; Wang, X.; Liu, J.; Sun, J. Effect of scanning strategies on the microstructure and mechanical behavior of 316L stainless steel fabricated by selective laser melting. *Mater. Sci. Eng. A* **2020**, *793*, 139879.

metals

MDPI

Article

Investigation of Microstructure and Mechanical Properties of the Repaired Precipitation-Strengthened Ni-Based Superalloy via Laser Melting Deposition

Wengao Yan [1,2,3,4], Beirao Xue [1,2,3,4], Jinjun Li [1,2,3], Minghuang Zhao [1,2,3] and Xiangde Bian [1,2,3,*]

[1] Advanced Gas Turbine Laboratory, Institute of Engineering Thermophysics, Chinese Academy of Sciences, Beijing 100190, China; 18674378306@163.com (W.Y.); xuebeirao@iet.cn (B.X.); lijinjun@iet.cn (J.L.); zhaominghuang@iet.cn (M.Z.)

[2] Key Laboratory of Advanced Energy and Power, Institute of Engineering Thermophysics, Chinese Academy of Sciences, Beijing 100190, China

[3] National Key Laboratory of Science and Technology on Advanced Light-duty Gas-Turbine, Chinese Academy of Sciences, Beijing 100190, China

[4] University of Chinese Academy of Sciences, Beijing 100049, China

* Correspondence: bianxiangde@iet.cn; Tel.: +86-156-5079-8558

Abstract: In this study, a typical γ' phase precipitation-strengthened Ni-based superalloy DZ411 was repaired using an LMD-based repairing technique with an IN738LC superalloy, and crack-free samples were acquired. The mechanical properties and microstructure of different areas inside the repair sample were investigated, including the IN738LC deposit, the DZ411 substrate, and the interface between these two parts. The differences in mechanical properties between different areas were explained via analyzing fractography and KAM maps. It was found that the coarse carbides of the DZ411 substrate might lead to rapid cracking of grain boundaries, resulting in the worst mechanical properties of the DZ411 substrate. The IN738LC deposit demonstrated significantly superior mechanical properties in comparison to the DZ411 substrate. Its tensile strength exceeded that of the substrate by over 250 MPa, while its relative elongation after fracture was twice as great as that of the substrate. The excellent mechanical properties of the IN738LC deposit could be attributed to its fine microstructure, which resisted rapid cracking and generated a large number of GNDs during the plastic deformation process. For the interface between the deposit and substrate, although its hardness before the tensile test was low, it could also generate many GNDs during the plastic deformation process, hence exhibiting commendable mechanical properties. The research results show that using an LMD-based repairing technique with IN738LC superalloy to repair γ' phase precipitation-strengthened Ni-based superalloy DZ411 is a feasible solution.

Keywords: Ni-based superalloy; repair; microstructure; tensile behavior; work hardening

1. Introduction

Precipitation-strengthened Ni-based superalloys are frequently used as turbine blade material in gas turbines, which operate in high-temperature, high-pressure, and high-speed environments, thereby making them susceptible to wear, cracking, and other damages [1–3]. The laser melting deposition (LMD) technique is an advanced technique that utilizes laser energy to melt metal powder, enabling the gradual formation of components via layer-by-layer stacking [4,5]. This technique provides the advantages of low heat input and high material availability, as well as the ability to ignore the complexity of structure, which makes it well-suited for repairing damaged blades [6,7].

However, achieving the successful reparation of precipitation-strengthened Ni-based superalloys is a considerable challenge. The primary factor contributing to this challenge is the poor weldability of the most precipitation-strengthened Ni-based superalloys, which leads to solidification cracks [8,9], liquation cracking [10], or some other defects [11] in the alloys during the repair procedure. Researchers have tried many methods to reduce the defect sensitivity of precipitation-strengthened Ni-based superalloys during LMD processing, including heating the substrate [10,12] and reducing heat input [13,14]. Despite extensive research efforts, the manufacturing of large-sized precipitation-strengthened Ni-based superalloys utilizing LMD remains difficult primarily because of the significant residual thermal stress inherent in the LMD process [14,15].

Due to the size limitation caused by residual thermal stress, there are only a few investigations on the mechanical properties of precipitation-strengthened Ni-based superalloys repaired or directly manufactured by LMD [16–18]. Xu et al. [16] manufactured the IN738LC superalloy using LMD, which had higher tensile strength compared with the as-cast IN738LC superalloy. Yu et al. [17] added pure Hf powder during the manufacture of IN738LC superalloy by selective laser melting, which successfully inhibited cracks and improved the mechanical properties of the alloy. Ci et al. [18] repaired DD32 single-crystal superalloy and investigated the relationship between the tensile behavior of the repaired DD32 superalloy and its microstructure. Within these small amounts of existing research, the mechanical properties of the interface between the deposit and substrate were frequently not investigated. For repaired superalloys, the mechanical properties of the interface are also important.

The heterogeneous interface often exhibits considerable compositional and microstructural gradients, resulting in distinctive mechanical properties. Fang et al. [19] designed a dual heterostructure brass that overcame the strength–ductility trade-off and possessed a superior combination of strength and ductility. This comprehensive strengthening effect was attributed to the generation of geometrically necessary dislocations (GNDs) at the heterogeneous interface during plastic deformation [20]. Tan et al. [21] also found that the gradient microstructures in additive-manufactured Ti–6Al–4V could enhance elongation by suppressing strain localization. However, the heterogeneous interface may not always strengthen the alloys. For example, during the melting joining process of two alloys with different compositions, brittle intermetallic compounds might be generated at the interface position, which reduces the interface performance and even induces cracks [22]. Therefore, it is necessary to investigate the mechanical properties of the interface of the repaired superalloy.

In this investigation, we repaired the DZ411 superalloy via LMD with IN738LC superalloy as the filler material, and some crack-free samples were obtained. The mechanical properties of the deposit, the substrate, and the interface between them were next measured, and an investigation was conducted to figure out the factors contributing to the variations in their mechanical properties.

2. Material and Methods

2.1. Materials

DZ411 superalloy is a typical γ' phase precipitation-strengthened Ni-based superalloy that is extensively utilized in turbine blades [23]. This alloy is also susceptible to damage, such as cracks, but due to its high Al and Ti content, it is difficult to use this alloy to repair itself directly [24]. The LMD technique utilized for the IN738LC superalloy has reached a considerable level of development [16,17]. In addition, the chemical composition and physico-thermal properties [25] of IN738LC superalloys are very similar to those of DZ411 superalloys, as shown in Tables 1 and 2. Therefore, it is a feasible method to repair the DZ411 superalloy by using the IN738LC superalloy as the filling material.

Table 1. Chemical composition of IN-738 superalloy and DZ411 superalloy.

Alloy	Ni	Cr	Co	W	Mo	Ti	Al	Nb	Ta	C	Si
IN738LC	bal	16.07	8.6	2.56	1.83	3.36	3.48	0.83	1.82	0.12	0.05
DZ411	bal	13.78	9.64	3.98	1.40	5.06	3.22	–	2.82	0.10	0.012

Table 2. Physico-thermal properties of IN-738 superalloy and DZ411 superalloy [25].

Alloy	Heat Conductivity [W/(m·°C), 1000 °C]	Specific Heat Capacity [J/(kg·°C), 1000 °C]	Thermal Expansivity [10^6 °C^{-1}), 20–1000 °C]
IN738LC	22.1	582	16.1
DZ411	22.5	575	15.9

2.2. Repair Process

The schematic diagram of the repair procedure is shown in Figure 1a. In this study, a carbon dioxide laser CP4000 was employed as a heat source. Although the heat input of the LMD to the substrate was relatively low, a pulsed laser was employed as the heat source to further minimize the heat input and limit the tendency of the sample to crack during the repair process [7]. The specific laser parameters were as follows: a power of 2000 W, a laser spot diameter of 2 mm, a laser spot interval of 0.7 mm, a pulse width of 0.16 s, and a pulse interval of 0.3 s.

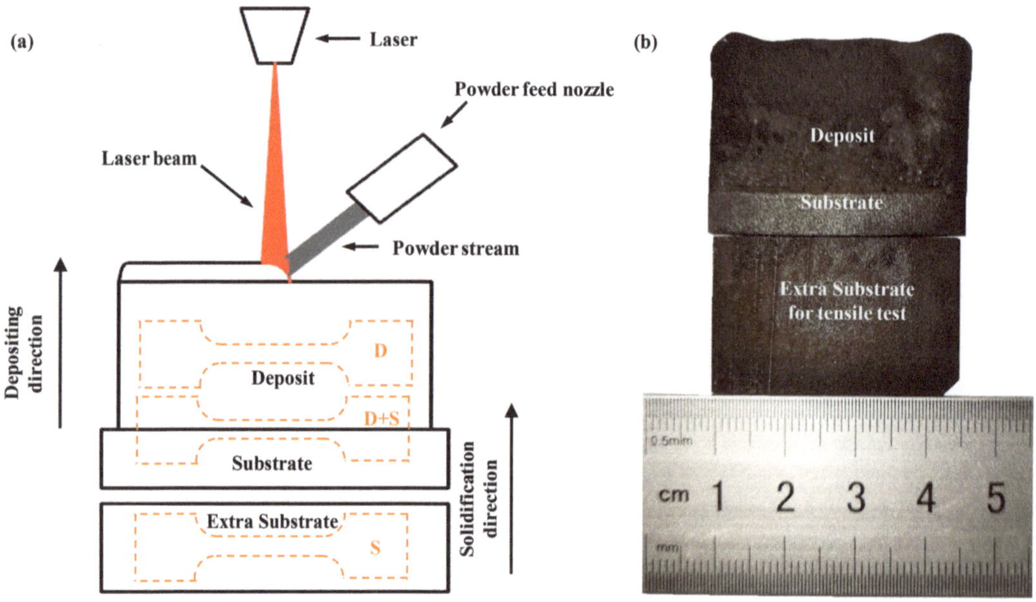

Figure 1. (**a**) Schematic diagram of the repairing process and the situation of the tensile samples; (**b**) typical photograph of repaired sample and extra substrate plate.

In the repairing process, the substrates were 3 mm thick DZ411 superalloy plates, which were manufactured by directional solidification method and subjected to the standard heat treatment of 1225 °C 2 h/AC + 1120 °C 2 h/AC + 850 °C 24 h/AC. The IN738LC superalloy powder (the powder morphology is shown in Figure 2) was fed directly into the molten pool by a nozzle using coaxial powder feeding technology. After melting, it was deposited layer by layer on the substrates, with the depositing direction aligning with the solidification direction of the substrates. After the repairing process, samples were subjected to a post-repairing heat treatment of 1120 °C 2 h/AC + 850 °C 24 h/AC, which referred to the standard heat treatment of IN738LC [26]. To compare the mechanical properties between the deposit and substrate, extra substrate plates were also subjected to post-repairing heat treatment. The typical photograph of the repaired sample and extra substrate plate is shown in Figure 1b. No cracks were found in all the repaired samples.

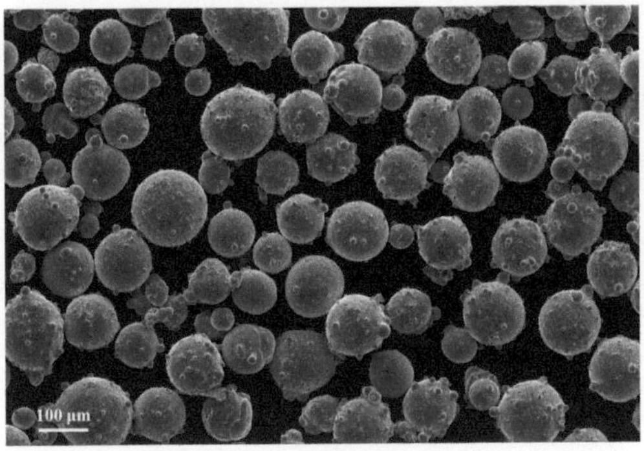

Figure 2. The morphology of IN738LC powder.

2.3. Microstructure Characterization and Tensile Test

To characterize the microstructure, the samples were polished and etched using an acid reagent (2.5 g $CuCl_2$, 50 mL HCl, and 50 mL H_2O). The microstructure of the repaired samples was characterized via optical microscopy (OM, LEICA-DM4000, Bruker, Billerica, MA, USA) and scanning electron microscopy (SEM, ZEISS-Gemini 300, ZEISS Group, Oberkochen, Germany).

In order to investigate the mechanical properties of each area of the repaired sample, tensile samples were taken from each area, with the main axis of the tensile sample being perpendicular to the depositing direction, as shown in Figure 1a. Tensile sample D is composed entirely of the IN738LC deposit; tensile sample S is composed entirely of DZ411 substrate; and tensile sample D + S is composed of both deposit and substrate. Three tensile tests were conducted on each type of tensile sample.

In addition, in order to analyze the changes in mechanical properties of local areas of the sample during the tensile test, the microhardness of the fractured samples was measured. The distribution of microhardness sampling points is depicted in Figure 3. These hardness values were compared to the undeformed repaired sample at the corresponding position to reflect the change in hardness between before and after the tensile test. An HAZ-1000 semiautomatic Viker tester was used for the microhardness test, and the test parameters were as follows: dwell time of 10 s and test load of 500 g.

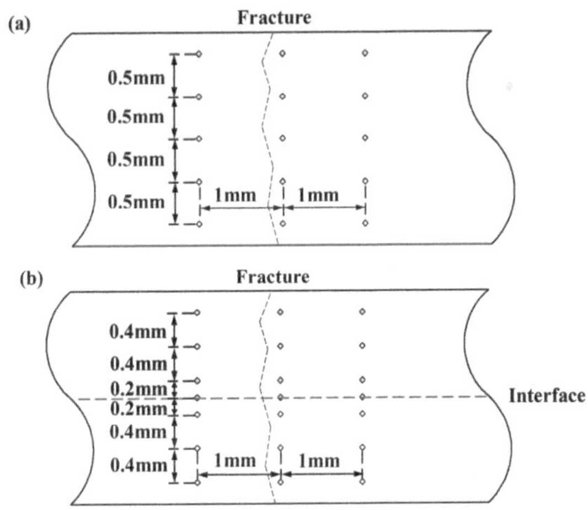

Figure 3. Schematic diagrams of the distribution of microhardness sampling points of (**a**) the tensile section of fractured D sample and S sample and (**b**) the tensile section of fractured D + S sample.

3. Results

3.1. Microstructure

The microstructure of the repaired sample is shown via OM and SEM in Figure 4. Apart from a tiny number of pores in the deposit, no additional visible defects were discovered in the repaired sample. The growth direction of the dendrites in the substrate stayed parallel to the solidification direction. Even though the dendritic width of the deposit was considerably lower than that of the substrate, most dendrite development directions were still inherited from the substrate. The refinement of dendrite width was a result of the quick cooling rate during the LMD process [8], and the consistency of growth direction was a result of the significant temperature gradient along the depositing direction during the LMD process [9]. A few equiaxed grains could also be observed near the interface, which might be formed by recrystallization.

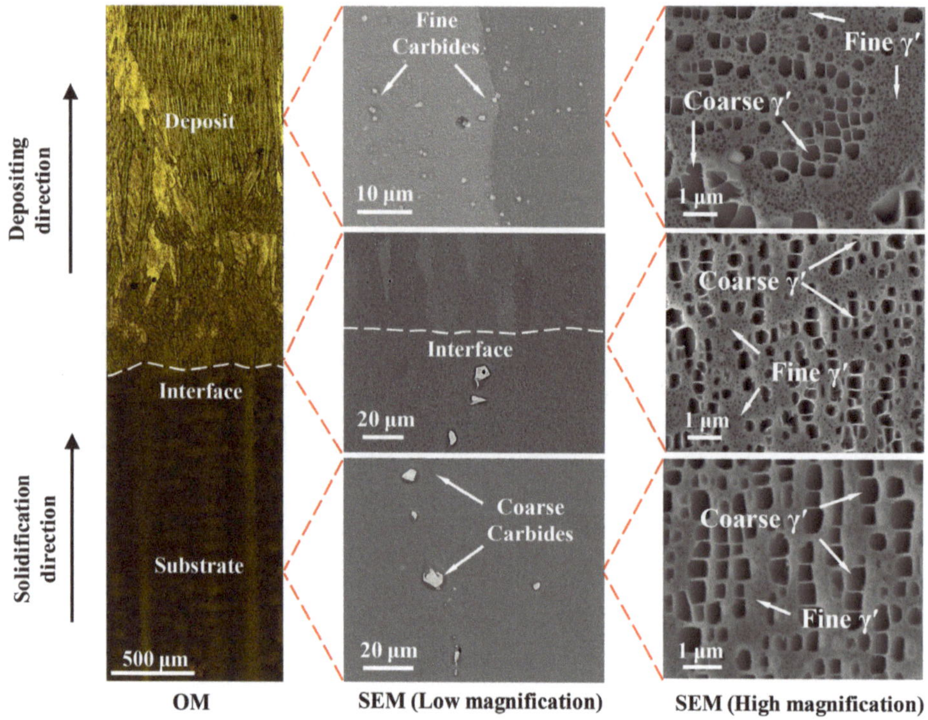

Figure 4. Microstructure of repaired sample.

The SEM images reveal that both the deposit and substrate exhibited a typical microstructure consisting of the γ phase as the matrix and the γ' phase as the predominant strengthening phase [27,28]. All areas contained a small number of carbides, and no intermetallic compounds were found to be generated at the interface. Low-magnification SEM images demonstrate a considerable difference in carbide morphology among different areas. The carbides in the deposit and interface were primarily tiny and possessed either a round or square shape, with an average diameter of less than 0.5 μm. However, the carbides in the substrate were significantly coarser and had an irregular shape, with an average diameter of more than 4.0 μm. This discrepancy in the morphology of carbides between the two areas was due to the different cooling rates during the manufacturing process of the two areas [10]. Compared to the substrate, during the manufacturing process of the deposit, the cooling rate was faster, the degree of element segregation was lower, and the morphology of the carbide was finer and more uniform. High-magnification SEM images reveal the bimodal shape of the γ' phase in each area of the repaired sample. The size of the fine γ' phase was comparable across all areas, measuring approximately 0.06 μm in diameter. The coarse γ' phase exhibited a consistent and progressive increase in size from the deposit to the substrate, with a side length of 0.28 μm in the deposit, a side length of 0.31 μm at the interface, and a side length of 0.38 μm in the substrate. According to Jackson and Reed [29], the coarse γ' phase was the phase that was not entirely dissolved in the solid solution treatment, whereas the fine γ' phase was the phase formed during the cooling process following the solid solution heat treatment. The substrate had undergone heat treatment before and after the repair procedure, so its coarse γ' phase had a longer growth time, resulting in a larger size and higher volume fraction of the coarse γ' phase [11].

3.2. Mechanical Properties

The stress–strain diagram and typical photograph of three types of tensile samples are shown in Figure 5. Due to the square shape of the tensile section of the samples, there was stress concentration at the corners of the sample. During the tensile testing, when the samples experienced pronounced localized plastic deformation, they might fracture rapidly from their corners. Hence, the sample photographs revealed a limited extent of plastic deformation, and the stress–strain curve for the period following necking was entirely absent.

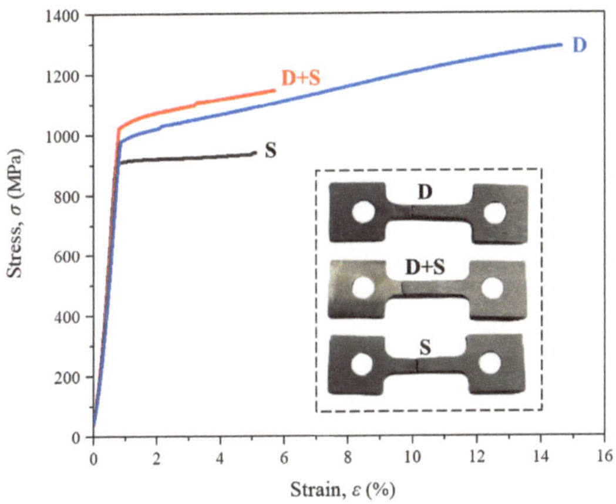

Figure 5. Stress–strain diagram and typical photograph of three types of tensile samples.

The tensile test results are shown in Table 3. In terms of ultimate tensile strength, it could be observed that D samples possessed the highest strength, while S samples possessed the lowest strength, with a notable disparity that exceeded 250 MPa between the two. The elongation values for both the S samples and the D + S samples were very similar, with both being about 6%. However, the D samples possessed an elongation value approximately twice as large as the previous two samples. The tensile test results indicated that the deposit had better mechanical properties than the substrate. Additionally, it could be noted that the D + S samples possessed the highest yield strength. According to the Hall–Petch relation [30], the large area grain boundary of fine equiaxed grains could hinder dislocation slip and increase the required driving force for dislocation slip, thus improving the yield strength of the alloy. Therefore, the high yield strength of the D + S samples might be attributed to the fine recrystallized equiaxed grains at the interface. Based on Table 3, it could also be found that the disparity in the ultimate tensile strength was much more significant than the disparity in the yield strength among the three types of samples. This indicated that the three types of samples exhibited a considerable difference in the work-hardening effect. The work-hardening effect of the D sample was the best, followed by that of the D + S samples, and that of the S samples was the worst.

Table 3. Tensile test results of three types of tensile samples.

Samples	Ultimate Tensile Strength (MPa)	Yield Strength (MPa)	Elongation (%)
D	942 ± 6	905 ± 11	5.8 ± 0.3
D + S	1102 ± 55	986 ± 40	6.0 ± 0.7
S	1240 ± 45	957 ± 37	10.7 ± 2.0

Figure 6 demonstrates the change in average hardness values of samples before and after the tensile test. The hardness values of the interface and the 0.2 mm area near the interface in the D + S sample are considered as the hardness of the interface. The hardness values of the remaining sampling points are defined as the hardness values of the deposit and the substrate in the D + S sample. Before the tensile test, it was seen that the hardness at the interface exhibited the lowest value, mostly because the recrystallized grains near the interface released internal stress, hence decreasing their hardness values. After being away from the recrystallization position, the hardness values increased. After the tensile test, the hardness values of all the samples increased, and there was a significant disparity in the hardness increment between different samples. Furthermore, the disparity in the hardness increment for each sample surpassed the disparity in hardness before the tensile test, indicating the disparity in hardness increment as the determining factor for the disparity in hardness after the tensile test. The observed increase in hardness could be attributed to the work-hardening effect of the samples, and the increment in hardness could serve as an indicator of the work-hardening effect [31]. In Figure 6, the hardness values of the D sample before the tensile test were comparable to those of the S sample, but the hardness increment of the D sample was greater than that of the S sample, resulting in the D sample having a higher hardness value than the S sample. This demonstrated that the work-hardening effect of the D sample was superior to that of the S sample, which was consistent with the analysis of the tensile test result. In the D + S sample, the deformation degree of each region was identical, and the hardness increment of the deposit was greater than that of the substrate, indicating that the work-hardening efficiency of the deposit was greater than that of the substrate. It was worth noting that the hardness of the interface in the D + S sample also increased greatly, indicating that the work-hardening efficiency of the interface was also high. Due to the high work-hardening efficiency of the deposit and interface in the D + S sample, the D + S sample could have a higher hardness value than the S sample under a similar degree of plastic deformation.

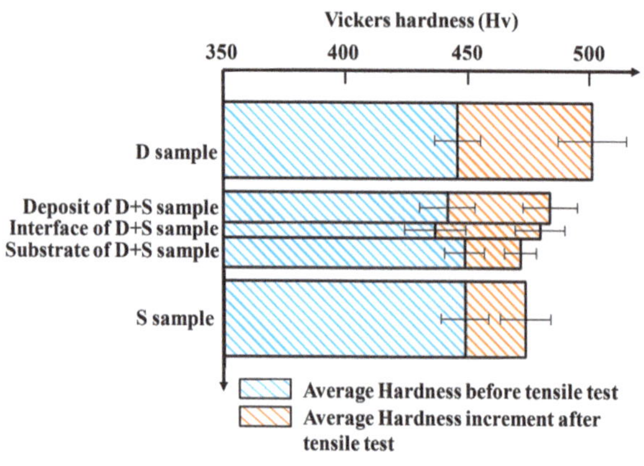

Figure 6. The change in average hardness of D sample, D + S sample, and S sample.

4. Discussion

The fractography of the three types of tensile samples is used to aid in the analysis of the mechanical property differences among the three, as shown in Figure 7. The fracture morphology of the S sample revealed that the S sample fracture mode was a brittle fracture mode that cracks along the grain boundary. On the fracture surface, there were numerous holes created by the carbide separated from the grain boundary, and under magnification, the remaining broken carbide could be observed. This phenomenon demonstrated that the coarse and irregularly shaped carbides in the S sample might result in mechanical property deterioration in the grain boundary. During the tensile test, micropores might develop at the boundary of these carbides, or generate from the directly broken carbides, which grew and finally linked to form cracks [32]. These cracks resulted in the rapid fracture of the S sample, which was the reason for the low relative elongation of the S sample.

The fracture morphology of the D sample indicated that it had a mixed fracture mode (Figure 7b). On the fracture surface of the D sample, two morphologies could be observed. One was a grain boundary cracking surface similar to that of the S sample but with a considerably smaller area, which might be attributable to the reduced grain size of the D sample. The other fracture morphology was a typical equiaxed dimple fracture, with tiny carbide particles in the middle of the equiaxed dimples. The size of carbides in the D sample was significantly smaller than that in the S sample, and the shape was more regular, resulting in less deterioration of the mechanical properties of the D sample. Hence, the superior ductility of the D sample might be a result of its fine grain and fine, uniform carbides.

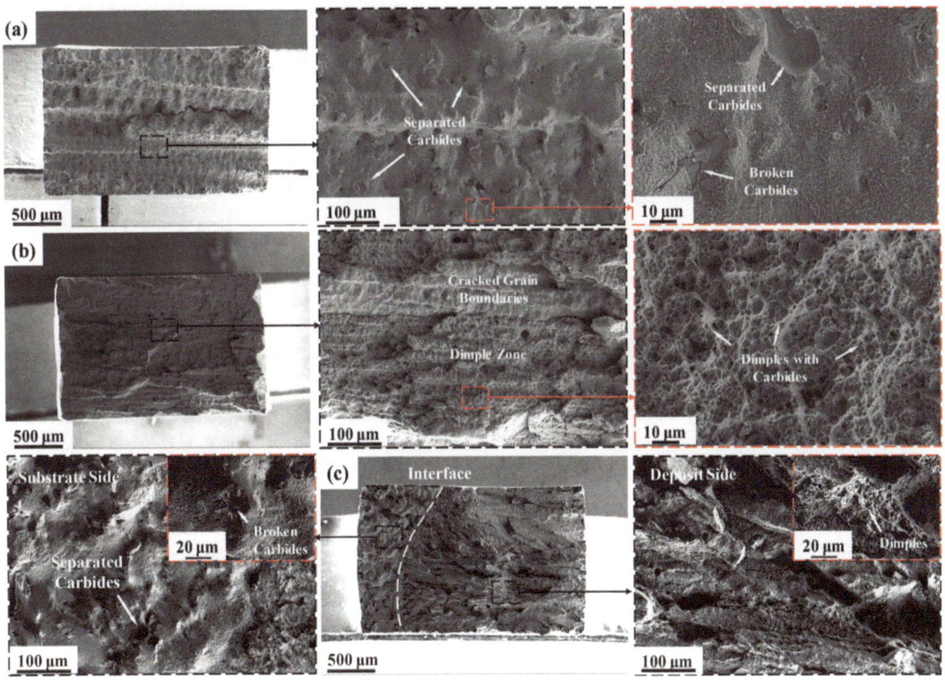

Figure 7. Fractography of three types of tensile samples: (**a**) S sample, (**b**) D sample, and (**c**) D + S sample.

The D + S sample had a more complicated fracture morphology. As shown in Figure 7c, the fracture of the D + S sample had two morphologies, which were found on the substrate side and the deposit side, respectively. The interface could clearly separate the two morphologies. The morphology of the substrate side was quite similar to that of the S sample, both of which were cracks along grain boundaries, and there were holes left by separated carbides and broken carbides on the surface. On the deposit side, a great number of tearing edges formed from the interface were spread, and there were still some dimples on the surface of the tearing edges. This morphology showed that the crack initiation of the D + S sample was similar to that of the S sample, that is, the cracks formed by the broken coarse carbides or the separation of carbides from the matrix. Therefore, D + S samples and S samples exhibited similar relative elongation.

The work-hardening effect of alloys is related to the density of geometrically necessary dislocations (GNDs) generated during plastic deformation [33]. EBSD kernel average misorientation (KAM) maps can illustrate the distribution of the density of GNDs in alloys. In the KAM map, the KAM values represent the local misorientation, which has a positive correlation with the density of GNDs [34]. Figure 8 shows the KAM maps of the three types of fractured samples near the fractured surface, and the thin white solid lines in Figure 8 represent the grain boundaries identified by the software. Figure 8a shows that the fractured D sample had the highest density of GNDs. This phenomenon might be attributed to the largest plastic deformation experienced by the D sample, leading to the generation of a large number of GNDs. In Figure 8b, for the D + S sample, the interface and nearby recrystallized grains (RG) had the highest KAM value. Due to the inconsistent mechanical properties of the deposit and the substrate in the D + S sample, there was a maximum strain gradient at the interface during the deformation of the D + S sample, promoting the generation of GNDs [35]. In addition, the grain size of the RG at the interface was very small, hence inhibiting the dislocation slip efficiently and hindering the dislocation annihilation during the plastic deformation process. Therefore, the interface and nearby RG had the maximal density of GNDs in the D + S sample after the tensile test, improving the hardness of the interface and its surrounding positions. It can also be noted that the deposit in the D + S sample had a higher density of GNDs than the substrate in the D + S sample, which might be because the microstructure of the deposit in the D + S sample was more refined than the substrate in the D + S sample. In Figure 8c, dislocations were primarily concentrated near carbides in the S sample. This phenomenon demonstrated that the carbides in the S sample had the ability to inhibit dislocation slipping. However, due to their large size and irregular shape, these carbides were prone to breaking, and their adhesion to the metal matrix was poor. When dislocations piled up at the boundary of these carbides, the carbides might directly break or separate from the matrix, which led to rapid fracture. Therefore, the diminished mechanical properties of the S sample could be attributed to the coarse carbides in the sample.

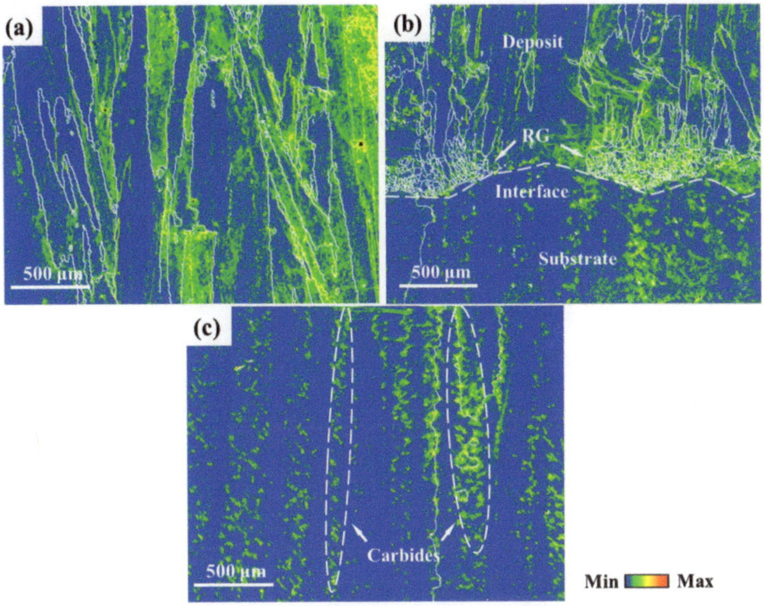

Figure 8. The KAM maps of (**a**) fractured D sample, (**b**) fractured D + S sample, and (**c**) fractured S sample.

5. Conclusions

Using an LMD-based repairing technique, crack-free repair samples with γ' phase precipitation-strengthened Ni-based superalloy DZ411 as substrate and IN738LC alloy as deposit were manufactured. The microstructure of the repaired samples was characterized, and the mechanical properties were measured. The main conclusions are as follows:

(1) Owing to the rapid cooling rate and high-temperature gradient during the LMD process, the microstructure in the deposit was evidently refined compared to that in the substrate, and the growth direction of the dendrite could be approximately consistent with the substrate.

(2) The S samples exhibited poor mechanical properties, as demonstrated by its ultimate tensile strength of 942 MPa and relative elongation of 5.8%. The reason for this is that large-sized carbides were prone to breaking, and their adhesion to the metal matrix was poor, which caused the S samples to fracture rapidly during the tensile test.

(3) Among the three types of samples, the D samples had the greatest mechanical properties, with an ultimate tensile strength of 1240 MPa and a relative elongation of 10.7%. These fine microstructures in the IN738LC deposit enhanced the ductility of the alloy, enabling it to generate more GNDs during the larger plastic deformation process, thereby strengthening the alloy.

(4) The crack initiation mechanism in the D + S sample exhibited similarity to that of the S sample, so the relative elongation of the two samples was equivalent, both below 6%. However, due to the greater work-hardening efficiency exhibited by the interface and deposit of the D + S sample, the D + S sample had greater strength and hardness after the tensile test than the S sample.

The research results show that using an LMD-based repairing technique with IN738LC superalloy to repair γ' phase precipitation-strengthened Ni-based superalloy DZ411 is a feasible solution and can provide some references for future research.

Author Contributions: W.Y.—Conceptualization, Investigation, Writing—original draft; B.X.—Data curation, Validation; J.L.—Validation; M.Z.—Investigation; X.B.—Conceptualization, Methodology, Writing—review and editing. All authors have read and agreed to the published version of the manuscript.

Funding: The work was financially supported by the National Science and Technology Major Project (J2019-IV-0006-0074).

Institutional Review Board Statement: Not applicable.

Informed Consent Statement: Not applicable.

Data Availability Statement: The data that support the findings of this study are available from the corresponding author, Xiang-de Bian, upon reasonable request.

Conflicts of Interest: The authors declare no conflict of interest.

References

1. Xu, R.; Li, Y.; Yu, H. Creep Behavior and Deformation Mechanism of a Third-Generation Single Crystal Ni-Based Superalloy at 980 °C. *Metals* **2023**, *13*, 1541. [CrossRef]
2. Yan, W.-G.; Zeng, W.; Man, J.-X.; Qiao, D.; Zhang, Z.-X.; Bian, X.-D. Microstructure and mechanical properties of a directionally solidified superalloy DZ411 plate sample repaired by laser melting deposition. *Mater. Sci. Eng. A* **2023**, *876*, 145141. [CrossRef]
3. Chen, J.; Chen, J.; Wang, Q.; Wu, Y.; Li, Q.; Xiao, C.; Hui, X. Mutation in TCP phases and superior stress rupture life led by W/Mo ratio in Ni-based single crystal superalloys. *Mater. Lett.* **2022**, *312*, 131656. [CrossRef]
4. Yang, Q.; Xu, Z.; Li, L.; Li, P. Microstructure and Performance Research on Ceramic-Enhanced Inconel 718 Matrix Composite Using Laser Additive Manufacturing. *Metals* **2023**, *13*, 1525. [CrossRef]
5. Wang, J.; Pan, Z.; Wang, Y.; Wang, L.; Su, L.; Cuiuri, D.; Zhao, Y.; Li, H. Evolution of crystallographic orientation, precipitation, phase transformation and mechanical properties realized by enhancing deposition current for dual-wire arc additive manufactured Ni-rich NiTi alloy. *Addit. Manuf.* **2020**, *34*, 101240. [CrossRef]
6. Li, L.; Deceuster, A.; Zhang, C. Effect of Process Parameters on Pulsed-Laser Repair of a Directionally Solidified Superalloy. *Met. Microstruct. Anal.* **2012**, *1*, 92–98. [CrossRef]
7. Moattari, M.; Shokrieh, M.M.; Moshayedi, H.; Kazempour-Liasi, H. Evaluations of residual stresses in repair welding of Ni-based IN939 superalloy. *J. Therm. Stress.* **2020**, *43*, 801–815. [CrossRef]
8. Rush, M.T.; Colegrove, P.A.; Zhang, Z.; Broad, D. Liquation and post-weld heat treatment cracking in Rene 80 laser repair welds. *J. Mater. Process. Technol.* **2012**, *212*, 188–197. [CrossRef]
9. Liu, G.; Du, D.; Wang, K.; Pu, Z.; Chang, B. Hot cracking behavior and mechanism of the IC10 directionally solidified superalloy during laser re-melting. *Vacuum* **2020**, *181*, 109563. [CrossRef]
10. Bidron, G.; Doghri, A.; Malot, T.; Fournier-Dit-Chabert, F.; Thomas, M.; Peyre, P. Reduction of the hot cracking sensitivity of CM-247LC superalloy processed by laser cladding using induction preheating. *J. Mater. Process. Technol.* **2019**, *277*, 116461. [CrossRef]
11. Griffiths, S.; Tabasi, H.G.; Ivas, T.; Maeder, X.; De Luca, A.; Zweiacker, K.; Wróbel, R.; Jhabvala, J.; Logé, R.; Leinenbach, C. Combining alloy and process modification for micro-crack mitigation in an additively manufactured Ni-base superalloy. *Addit. Manuf.* **2020**, *36*, 101443. [CrossRef]
12. Guo, C.; Li, G.; Li, S.; Hu, X.; Lu, H.; Li, X.; Xu, Z.; Chen, Y.; Li, Q.; Lu, J.; et al. Additive manufacturing of Ni-based superalloys: Residual stress, mechanisms of crack formation and strategies for crack inhibition. *Nano Mater. Sci.* **2023**, *5*, 53–77. [CrossRef]
13. Xu, J.; Ding, Y.; Gao, Y.; Wang, H.; Hu, Y.; Zhang, D. Grain refinement and crack inhibition of hard-to-weld Inconel 738 alloy by altering the scanning strategy during selective laser melting. *Mater. Des.* **2021**, *209*, 109940. [CrossRef]
14. Guo, C.; Zhou, Y.; Li, X.; Hu, X.; Xu, Z.; Dong, E.; Zhu, Q.; Ward, R.M. A comparing study of defect generation in IN738LC superalloy fabricated by laser powder bed fusion: Continuous-wave mode versus pulsed-wave mode. *J. Mater. Sci. Technol.* **2021**, *90*, 45–57. [CrossRef]
15. Carpenter, K.; Tabei, A. On Residual Stress Development, Prevention, and Compensation in Metal Additive Manufacturing. *Materials* **2020**, *13*, 255. [CrossRef]
16. Xu, J.; Lin, X.; Guo, P.; Yang, H.; Xue, L.; Huang, W. The microstructure evolution and strengthening mechanism of a γ'-strengthening superalloy prepared by induction-assisted laser solid forming. *J. Alloys Compd.* **2019**, *780*, 461–475. [CrossRef]
17. Yu, Z.; Guo, C.; Han, S.; Hu, X.; Cao, L.; Xu, Z.; Ding, H.; Zhu, Q. The effect of Hf on solidification cracking inhibition of IN738LC processed by Selective Laser Melting. *Mater. Sci. Eng. A* **2021**, *804*, 140733. [CrossRef]
18. Ci, S.; Liang, J.; Li, J.; Zhou, Y.; Sun, X. Microstructure and tensile properties of DD32 single crystal Ni-base superalloy repaired by laser metal forming. *J. Mater. Sci. Technol.* **2020**, *45*, 23–34. [CrossRef]
19. Fang, X.; Li, Z.; Wang, Y.; Ruiz, M.; Ma, X.; Wang, H.; Schoell, R.; Zheng, C.; Kaoumi, D.; Zhu, Y. Achieving high hetero-deformation induced (HDI) strengthening and hardening in brass by dual heterostructures. *J. Mater. Sci. Technol.* **2021**, *98*, 244–247. [CrossRef]

20. Wu, X.; Zhu, Y.; Lu, K. Ductility and strain hardening in gradient and lamellar structured materials. *Scr. Mater.* **2020**, *186*, 321–325. [CrossRef]
21. Tan, X.; Kok, Y.; Tan, Y.J.; Descoins, M.; Mangelinck, D.; Tor, S.B.; Leong, K.F.; Chua, C.K. Graded microstructure and mechanical properties of additive manufactured Ti–6Al–4V via electron beam melting. *Acta Mater.* **2015**, *97*, 1–16. [CrossRef]
22. Yuhua, C.; Yuqing, M.; Weiwei, L.; Peng, H. Investigation of welding crack in micro laser welded NiTiNb shape memory alloy and Ti6Al4V alloy dissimilar metals joints. *Opt. Laser Technol.* **2017**, *91*, 197–202. [CrossRef]
23. Chen, Y.; Yao, Z.; Dong, J.; Zhao, W.; Xiao, H.; Chen, X. Molecular dynamics simulation of the γ' phase deformation behaviour in nickel-based superalloys. *Mater. Sci. Technol.* **2022**, *38*, 1439–1450. [CrossRef]
24. Peng, P.; Lu, L.; Liu, Z.; Pei, X.; Gan, L.; Xu, Y.; Zhang, X.; Ma, Z.; Guo, M.; Liu, L. Investigation on the influence of Ta on the micro-structure evolution of Ni-based superalloy DZ411 during directional solidification, heat treatment, and long-term aging. *J. Alloys Compd.* **2022**, *920*, 165886. [CrossRef]
25. Academic Committee of the Superalloys CSM. *China Superalloys Handbook;* Standards Press of China: Beijing, China, 2012.
26. Xu, J.; Gruber, H.; Boyd, R.; Jiang, S.; Peng, R.L.; Moverare, J.J. On the strengthening and embrittlement mechanisms of an additively manufactured Nickel-base superalloy. *Materialia* **2020**, *10*, 100657. [CrossRef]
27. Holländer, D.; Kulawinski, D.; Thiele, M.; Damm, C.; Henkel, S.; Biermann, H.; Gampe, U. Investigation of isothermal and ther-mo-mechanical fatigue behavior of the nickel-base superalloy IN738LC using standardized and advanced test methods. *Mater. Sci. Eng. A* **2016**, *670*, 314–324. [CrossRef]
28. Wu, Y.; Chen, J.; Zhang, L.; Ji, J.; Wang, Q.; Zhang, S. Effect of boron on the structural stability, mechanical properties, and electronic structures of γ'-Ni3Al in TLP joints of nickel-based single-crystal alloys. *Mater. Today Commun.* **2022**, *31*, 103375. [CrossRef]
29. Jackson, M.; Reed, R. Heat treatment of UDIMET 720Li: The effect of microstructure on properties. *Mater. Sci. Eng. A* **1999**, *259*, 85–97. [CrossRef]
30. Guo, B.; Zhang, Y.; He, F.; Ma, J.; Li, J.; Wang, Z.; Wang, J.; Feng, J.; Wang, W.; Gao, L. Origins of the mechanical property heterogeneity in a hybrid additive manufactured Hastelloy X. *Mater. Sci. Eng. A* **2021**, *823*, 141716. [CrossRef]
31. Movahed, P.; Kolahgar, S.; Marashi, S.; Pouranvari, M.; Parvin, N. The effect of intercritical heat treatment temperature on the tensile properties and work hardening behavior of ferrite–martensite dual phase steel sheets. *Mater. Sci. Eng. A* **2009**, *518*, 1–6. [CrossRef]
32. Li, J.; Wang, H. Microstructure and mechanical properties of rapid directionally solidified Ni-base superalloy Rene'41 by laser melting deposition manufacturing. *Mater. Sci. Eng. A* **2010**, *527*, 4823–4829. [CrossRef]
33. Zhang, S.; Lin, X.; Wang, L.; Yu, X.; Hu, Y.; Yang, H.; Lei, L.; Huang, W. Strengthening mechanisms in selective laser-melted Inconel718 superalloy. *Mater. Sci. Eng. A* **2021**, *812*, 141145. [CrossRef]
34. Yan, Z.; Wang, D.; He, X.; Wang, W.; Zhang, H.; Dong, P.; Li, C.; Li, Y.; Zhou, J.; Liu, Z.; et al. Deformation behaviors and cyclic strength assessment of AZ31B magnesium alloy based on steady ratcheting effect. *Mater. Sci. Eng. A* **2018**, *723*, 212–220. [CrossRef]
35. Li, X.; Lu, L.; Li, J.; Zhang, X.; Gao, H. Mechanical properties and deformation mechanisms of gradient nanostructured metals and alloys. *Nat. Rev. Mater.* **2020**, *5*, 706–723. [CrossRef]

metals

MDPI

Article

Structure and Properties Evolution of AZhK Superalloy Prepared by Laser Powder Bed Fusion Combined with Hot Isostatic Pressing and Heat Treatment

Fedor A. Baskov [1,*], Zhanna A. Sentyurina [2], Pavel A. Loginov [1], Marina Ya. Bychkova [1], Ivan A. Logachev [1] and Evgeny A. Levashov [1]

[1] Scientific-Educational Center of SHS, National University of Science and Technology 'MISIS', Moscow 119049, Russia; pavel.loginov.misis@list.ru (P.A.L.); bychkova@shs.misis.ru (M.Y.B.); logachev.ia@misis.ru (I.A.L.); levashov@shs.misis.ru (E.A.L.)

[2] PJSC MMC 'Norilsk Nickel', Moscow 123100, Russia; sentyurinazha@nornik.ru

* Correspondence: baskov.fa@misis.ru

Abstract: The structure and properties of samples obtained by the laser powder bed fusion (LPBF) method from the AZhK alloy, intended for the manufacture of heavily loaded body parts with operating temperatures up to 800 °C, have been studied. The optimal mode of LPBF, ensuring the attainment of the minimal residual porosity of 0.02%, was identified for the superalloy AZhK. Additionally, the evolution of the microstructure of LPBF samples after hot isostatic pressing (HIP) and heat treatment (HT) was studied using optical microscopy (OM), scanning electron microscopy (SEM), and transmission electron microscopy (TEM). The macrostructure of LPBF samples is represented by columnar grains oriented in the direction of predominant heat dissipation, perpendicular to the build plate. At the microlevel, the structure consists of colonies of columnar dendrites. Nb_4AlC_3 and Nb_6C_4 carbides, as well as the Mo_2Hf Laves phase, are precipitated in the interdendritic region as a result of doping element segregation. The low strength of the LPBF samples (σ = 967 ± 10 MPa) is caused by the fact that there are no reinforcing particles and by high internal stress due to high crystallization speed. HIP and HT were found to have a favorable effect on the structure and properties of the LPBF samples. The post-treatment resulted in uniform distribution of γ'-phase precipitates sized up to 250 nm in the matrix bulk and carbides at grain boundaries, as well as Laves phase dissolution. Therefore, the strength characteristics were significantly improved: by 45% at room temperature and by 50% at elevated temperatures. High strength and ductility were attained (at 20 °C, σ^{20} = 1396 ± 22 MPa and δ = 19.0 ± 3.0 %; at 650 °C, σ^{650} = 1240 ± 25 MPa and δ = 15.8 ± 1.5%; at 750 °C, σ^{750} = 1085 ± 23 MPa and δ = 9.1 ± 2.3%). An ejector-type part was fabricated, and its geometric parameters coincided with those in the electronic models. After conducting computed tomography, it was found that there are no defects in the form of non-fusion and microcracks within the volume of the part; however, it was observed that the pore size is ≥20 μm.

Keywords: nickel-based superalloy; hot isostatic pressing (HIP); laser powder bed fusion (LPBF); transmission electron microscopy (TEM); mechanical properties

1. Introduction

Additive manufacturing (AM) is one of the most rapidly developing trends in the fabrication of parts from metallic materials [1–3]. The keen interest in AM is caused by the fact that it allows one to perform topology optimization of components, use fewer assembly units in the end product, reduce the weight of metallic structures and the production cycle time, as well as make manufacturing more flexible [4–6].

The fast pace of implementing AM into commercial production and integrating with it is observed for the engine-building and energy industries, where there is particular demand for geometrically complex parts made of expensive materials [7–10]. The overwhelming majority of studies in this field focus on developing the processes for fabricating products from superalloys, which are characterized by poor machinability [11,12].

Many parts can be manufactured from nickel superalloys using laser powder bed fusion (LPBF) technology, which ensures high accuracy of part building and better surface condition compared to other AM methods [13–15]. The past decade has witnessed a trend towards the expansion of the list of nickel alloys for the LPBF method. Alloys such as Inconel 718, Inconel 625, IN738LC, Hastelloy X, CM247LC, EP741NP, and some others are used in industry to different extents [15–23]. Further expansion of the range of alloy brands aims to improve manufacturability in order to suppress crack formation during the LPBF process and increase service temperature.

The AZhK alloy intended for fabricating heavy-duty casing parts operating at temperatures up to 800 °C was the study object of this paper. This alloy was originally developed to be used in net-shape powder metallurgy by hot isostatic pressing (HIP) technology, which is rather expensive because of the large number of process stages [24–27]. However, the alloy doping system ensures good weldability and manufacturability of the alloy, making it an excellent candidate for use in the LPBF method. The transition from the conventional technology to LPBF will make manufacturing of casing parts less time- and resource-consuming [28–32].

The aim of this study was to optimize laser powder bed fusion parameters as applied to the AZhK alloy using the fabrication of an ejector-type part as an example, as well as to study the effect of hot isostatic pressing and heat post-treatment on the structure and mechanical properties of LPBF samples.

2. Materials and Methods

2.1. Starting Materials

Powder of the AZhK alloy for LPBF, which was fabricated by the plasma rotating electrode process, was used in this study. AZhK is a Ni-Cr-Mo-Co superalloy containing significant amounts of Al and Nb. The typical chemical composition of AzhK is shown in Table 1.

The particle size of the powder varied from 20 to 63 μm, with an average particle size of 45 μm; the d_{10}, d_{50}, and d_{90} quantiles of distribution were 23.0, 36.2, and 54.1 μm, respectively. Powder particles were spherical in shape, without satellites and gas pores (Figure 1a). The internal structure of the powders was represented by fine dendrites; no closed gas pores were detected (Figure 1). The bulk density measured according to ISO 3923–1:2008 was 4.61 g/cm^3. The flowability of the powder determined in compliance with ASTM B213-17 was 14.4 s.

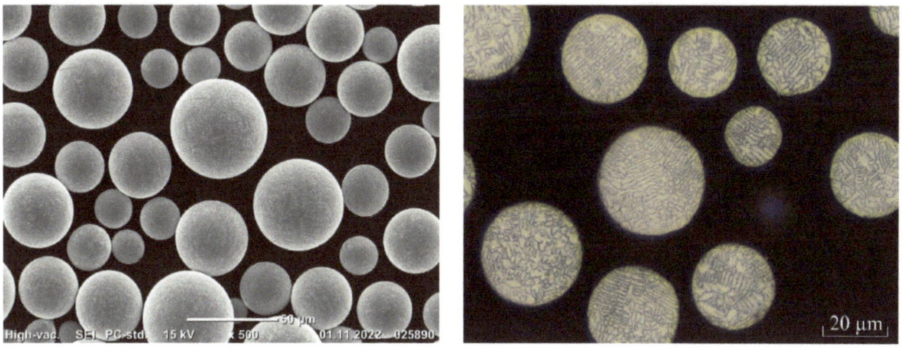

(a) (b)

Figure 1. The morphology (**a**) and microstructure (**b**) of the AZhK alloy powder.

Table 1. The chemical composition of the AZhK alloy.

Major Elements (Mass Fraction, wt. %)							
Ni	Cr	Mo	Co	Al	Nb	Hf	C
base	15.0–16.0	7.0–9.0	5.0–7.0	4.0–5.0	2.5–3.5	0.1–0.4	0.02–0.05
Impurities (mass fraction, wt. %)							
Si	Mn	Fe	P		S	O	N
0.5	0.5	0.5	up to 0.015		up to 0.009	<0.007	<0.002

2.2. The LPBF Process

Laser powder bed fusion of AZhK powder was conducted in an argon atmosphere on a TruPrint 1000 machine (Trumpf, Ditzingen, Germany) equipped with a fiber laser with a rated power of 200 W and a build platform 100 mm in diameter. In the parametric studies, we varied the scan speed V from 500 to 800 mm/s and the hatch distance x from 0.05 to 0.07 mm, while power P, powder layer height h, and laser beam spot diameter d were kept constant (P = 120 W, h = 0.02 μm, and d = 0.03 mm). The working platform was not heated. The combined effect of the technological parameters of the LPBF process on the characteristics of the material was assessed using a volumetric energy density, *VED* [J/mm^3]:

$$VED = \frac{P}{V \cdot h \cdot x} \tag{1}$$

where P is the laser power (W); V is the scan speed (mm/s); h is the powder layer height (mm); and x is the hatch distance (mm) [33–35].

Table 2 lists the analyzed laser powder bed fusion modes for building the samples and the respective *VED* values. The employed scanning strategy is schematically shown in Figure 2.

Table 2. The LPBF modes used in building the samples from AZhK powder.

No.	Mode				
	A	B	C	D	E
	VED, J/mm^3				
1	240.0	200.0	184.6	171.4	150.0
2	200.0	166.7	153.8	142.9	125.0
3	171.4	142.9	131.9	122.4	107.1

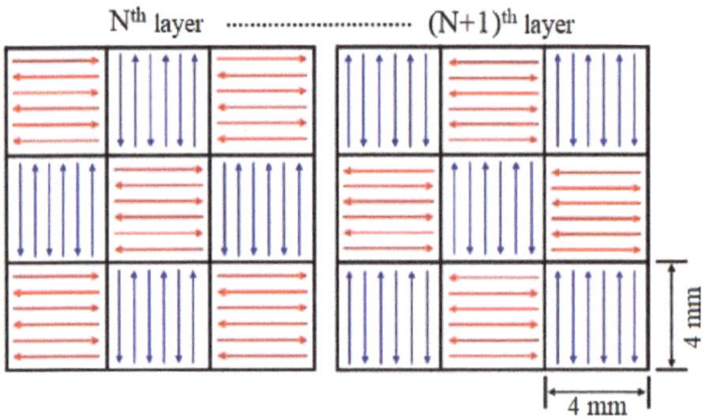

Figure 2. The chessboard scanning strategy.

The optimal LPBF mode was selected according to the results of structural studies performed for the samples sized 10 mm × 10 mm × 10 mm (Figure 3a) and prepared during the parametric studies in the modes summarized in Table 2. Vertical cylindrical samples for mechanical tests (Figure 3b) and the ejector-type part (Figure 3c) were fabricated in the optimal mode (E—125.0 J/mm^3). The cylindrical samples were 15 mm in diameter and 65 mm in height.

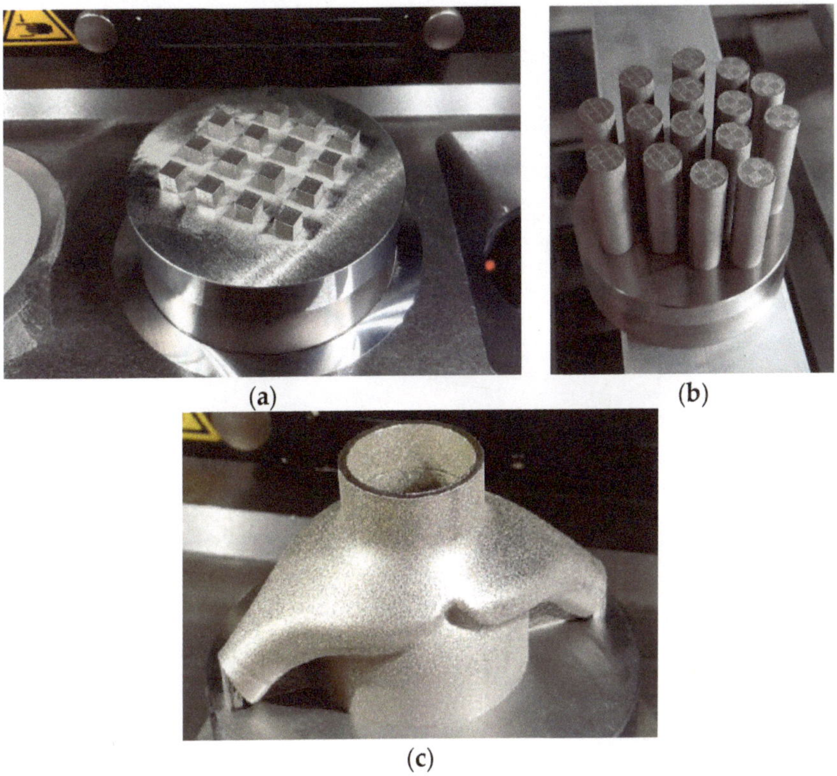

Figure 3. Appearance of the LPBF samples fabricated from the AZhK alloy for optimizing the LPBF mode (**a**) and for studying properties of the material (**b**), as well as the ejector-type part (**c**).

2.3. *Post-Treatment*

The samples produced by LPBF were subsequently subjected to various post-treatments: hot isostatic pressing (HIP), heat treatment (HT: solution + aging), and HIP + HT. HIP was carried out in an ABRA HIRP 10/26-200-2000 gasostat (ABRA AG, Winterthur, Switzerland) at 1160 °C and 150 MPa. Solution was performed at 1180 °C for 4 h. Aging was carried out at 820 °C for 16 h in a PL-10/16s thermal furnace (Russia). After quenching and aging, the samples were air-cooled.

2.4. *Characterization*

The structure and properties were studied for the following samples: LPBF samples; LPBF samples + HIP; LPBF samples + HT; LPBF samples + HIP + HT. The macrostructure and porosity of the samples were analyzed by optical microscopy (OM) using an AXIO Imager A1 reflection light microscope (CarlZeiss, Jena, Germany). The microstructure was investigated by scanning electron microscopy (SEM) on an S-3400N microscope (Hitachi High-Technologies Corporation, Tokyo, Japan). In order to detect the structural components, thin sections were etched using a solution of Marble's reagent (copper sulfate, 20 g;

hydrochloric acid, 100 mL; and water, 100 mL). High-resolution (HR) fine-structure study and phase identification were performed on a JEM-2100 transmission electron microscope (TEM) (Jeol, Tokyo, Japan). Foils for the HRTEM studies were prepared using a PHIPS II System dual-beam ion etching machine (Gatan, Inc., Pleasanton, CA, USA).

The hardness of the samples was measured on an HVS-50 Vickers hardness tester (Time Group Inc., Beijing, China) according to the State Standard GOST 2999 at a load of 10 kg. Tensile testing of the samples was carried out at room temperature in compliance with the State Standard GOST 1497 (ISO 6892), as well as at 650 and 750 °C in compliance with the State Standard GOST 9651 (ISO 783). The following characteristics were determined in the mechanical tests: ultimate tensile strength σ, yield point $\sigma_{0.2}$, and relative elongation δ. A Schenk–Trebel RMC-100 testing machine (Schenk, Darmstadt, Germany) and an Ulvac sinku Rico furnace (ULVAC Technologies, Methuen, MA, USA) were used for the mechanical tests.

The geometric dimensions of the ejector-type part made of AZhK alloy were controlled using a Kreon ACE Arm-7 portable measuring arm (Kreon Technologies, Limoges, France). Control over internal defects in the ejector-type part was performed using a Phenix V | tome | x M300 X-ray microfocus CT system (Waygate Technologies, Wunstorf, Germany) equipped with two X-ray tubes for regulation and conducting linear measurements with an accuracy of 4 μm + L (mm)/100. The ejector-type part was scanned in the following mode: voltage, 270 kV; current, 178 μA; voxel size, 51 μm; and total scan time, 74 min.

3. Results and Discussion

3.1. The Structural Features of the LPBF Samples

An analysis of the test samples prepared in the parametric studies revealed the following structure types:

- Structures with large circular pores, which have presumably resulted from boiling of the material (Figure 4a);
- Structures with micropores distributed over the entire sample volume (Figure 4b);
- Structures with isolated micropores (Figure 4c).

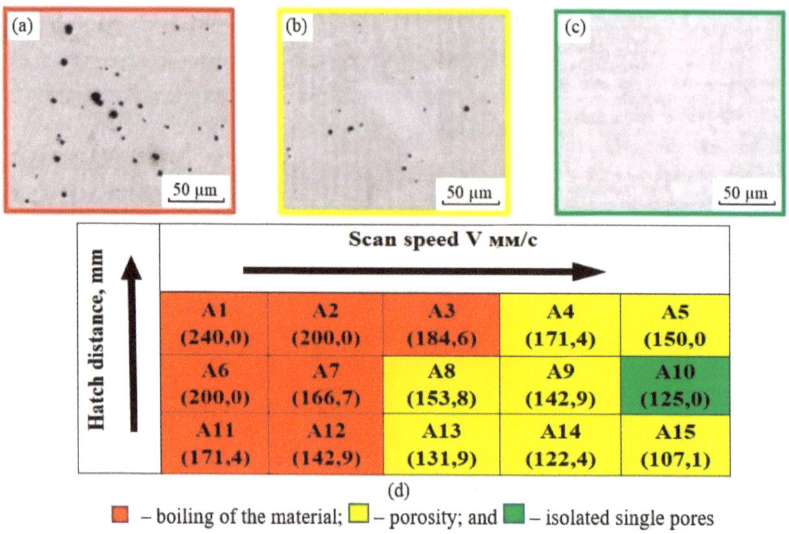

Figure 4. The characteristic types of the structure of the AZhK alloy (**a**–**c**) and the parametric study matrix for the LPBF process (**d**).

Figure 4d shows the experimental design matrix superimposed with the results of the structural studies. The top row of cells presents the notation of the modes (A1, A2, A3, etc.), while the bottom row of cells displays the porosity values calculated by analyzing the OM images. An analysis of the reported results confirms the hypothesis that the material boils at high energy inputs, resulting in the formation of large pores over the entire material volume. Porosity decreased as the energy input was reduced by increasing the scan speed. It is worth noting that the samples fabricated in the entire range of modes under study contained neither cracks nor incomplete fusion defects, thus proving the high manufacturability of the alloy during welding. The sample fabricated in mode E2 (E = 125.0 J/mm^3) contained the fewest defects. This mode was chosen to be the optimal one for further studies of the structure and mechanical characteristics.

Figure 5a,b shows the macrostructure of the LPBF samples in the longitudinal and transverse cross-sections. In the cross-section parallel to the XY build plane, one can clearly see laser tracks (scan paths), while arcuate molten pool boundaries and columnar grains sized up to 200 μm were observed in the YZ build plane. The structure of the resulting samples was characterized by the minimal number of defects and residual porosity of <0.02%. The single pores were sized 1–5 μm. No structural defects such as cracks or incomplete fusion regions were detected.

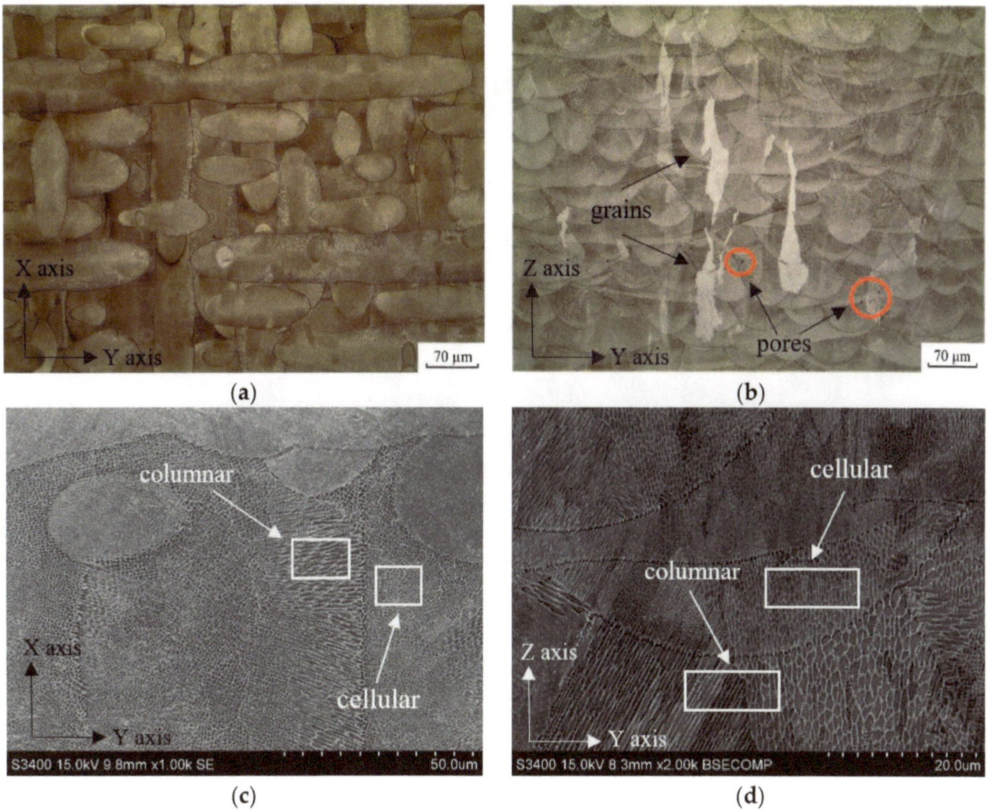

Figure 5. The OM (**a**,**b**) and SEM (**c**,**d**) images of the microstructure of the LPBF samples.

Figure 5c,d shows the results of SEM structural studies. The presented SEM images demonstrate that the structure of the LPBF samples consisted of cellular and columnar primary dendrites. The columnar microstructure was formed at the molten pool boundary in the form of single primary dendrites, where they grew in the direction of heat dissipation during subsequent crystallization, which was accompanied by the formation of oriented colonies. Epitaxial growth of dendritic colonies through the molten pool boundary became possible as the next layer was welded on, resulting in the formation of columnar grains (Figure 5b). Suppression of the epitaxial mechanism was predominantly observed for this alloy due to the switching of the growth direction of columnar primary dendrites at the molten bath boundary by 90°, so that it had a cellular microstructure (Figure 5c).

Figure 6a shows a TEM image of the fine structure of the LPBF sample made of the AZhK alloy; Figure 6b presents an electron diffraction pattern for a dendritic cell oriented so that the [0–11] axis lay parallel to the optical axis of the microscope and corresponded to the Ni-based matrix (the austenite γ-phase) with the calculated fcc lattice parameter of 3.5174 Å (the tabular value was 3.5238 Å). The structure of the synthesized material was characterized by high density of dislocations predominantly aggregated along the boundaries of the dendritic cells of the γ-phase. The cell size varied from 0.5 to 1.0 μm. Dislocation aggregates were indicative of high internal stress in the alloy after LPBF. Therefore, relaxation annealing was needed before the parts were cut off from the build plate when building geometrically complex and thin-walled items, as it prevented geometry distortion caused by their deformation.

(**a**) (**b**)

Figure 6. A TEM image of the structure of the LPBF sample (**a**) and an electron diffraction pattern (**b**) recorded for the analyzed region.

To assess the segregation of heavy alloying elements, local EDX analysis was carried out at points corresponding to the matrix (spectra 2, 3, 8, 14), the boundaries of dendritic grains (spectra 4, 5, 6, 12), and the precipitated particles (spectra 1, 7, 9, 10, 11, 13) presented in Figure 7a,b. The results of the local EDX are presented in Table 3.

The presented results indicate that the boundaries of the dendritic cells are mainly enriched in niobium, while the particles in the interdendritic space are enriched in elements such as molybdenum, niobium, and hafnium, and in one case, a particle with carbon and increased aluminum content is observed (spectrum 7).

The phases detected in the interdendritic space were identified by HRTEM. Figure 8 shows images of the particles detected in the interdendritic space and the respective Fourier transform (FFT) images. An analysis of FFT revealed that there were two carbides ($Nb_6C_{4.87}$ and Nb_4AlC_3) and the Laves phase (Mo_2Hf). Table 4 summarizes the results of the calculations. These phases were brittle and unstable; they were undesirable because they deteriorated mechanical properties of the material [36–41]. In order to get rid of these phases, we suggest conducting post-treatment procedures (HIP and HT) [22,23,38,42].

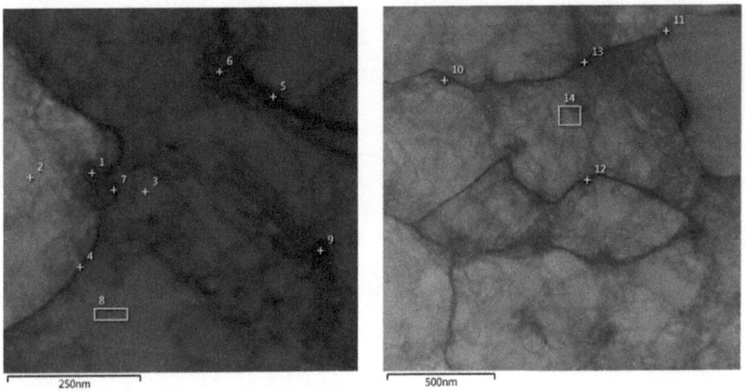

(a) (b)

Figure 7. Images of the structure of the LPBF sample from the AZhK alloy at various magnifications, indicating the regions of the local EDX. (**a**) area of study 1, (**b**) area of study 2.

Table 3. The results of the local EDX of the LPBF sample structure.

Spectrum	Content of Elements, wt. %							
	Ni	Co	Cr	Al	Mo	Nb	Hf	C
1	42.40	6.21	13.86	0.83	17.81	12.38	6.51	-
2	64.56	6.71	15.86	3.06	7.49	2.32	-	-
3	64.73	6.65	15.94	3.35	6.87	2.46	-	-
4	64.07	4.51	13.10	2.91	7.61	7.80	-	-
5	63.80	5.07	13.42	2.90	7.20	7.61	-	-
6	62.90	4.98	13.00	2.73	8.25	8.14	-	-
7	37.62	3.92	10.03	5.33	11.78	14.36	3.75	13.21
8	63.50	6.47	15.83	3.48	8.42	2.30	-	-
9	65.38	6.56	15.58	2.82	7.32	2.34	-	-
10	47.22	4.57	13.22	1.61	17.65	9.13	6.60	-
11	46.01	4.45	13.95	1.27	17.25	11.04	6.03	-
12	64.37	4.57	12.48	2.48	7.81	8.29	-	-
13	45.77	4.64	13.43	1.84	16.34	12.03	5.95	-
14	64.41	6.11	16.15	3.14	7.76	2.43	-	-

Table 4. The phase composition of the precipitates detected by TEM at the periphery of dendritic cells of the LPBF samples fabricated from the AZhK alloy.

Phase	Figure	Zone Axis *	Space Group	Lattice Type	Lattice Parameters	Tabular Values **
$Nb_6C_{4.87}$	45	[2-1-10]	P31	hcp	$a = 5.507$ $c = 14.946$	$a = 5.464$ $c = 15.422$
Nb_4AlC_3	44	[2-1-10]	P63/mmc	hcp	$a = 3.1155$ $c = 23.1197$	$a = 3.1296$ $c = 24.1208$
Mo_2Hf	42	[01-1]	Fd-3m	cubic	$a = 7.587$	$a = 7.545$

* Axis of the zone along which the crystallite of the detected phase was oriented. ** Tabular lattice parameters.

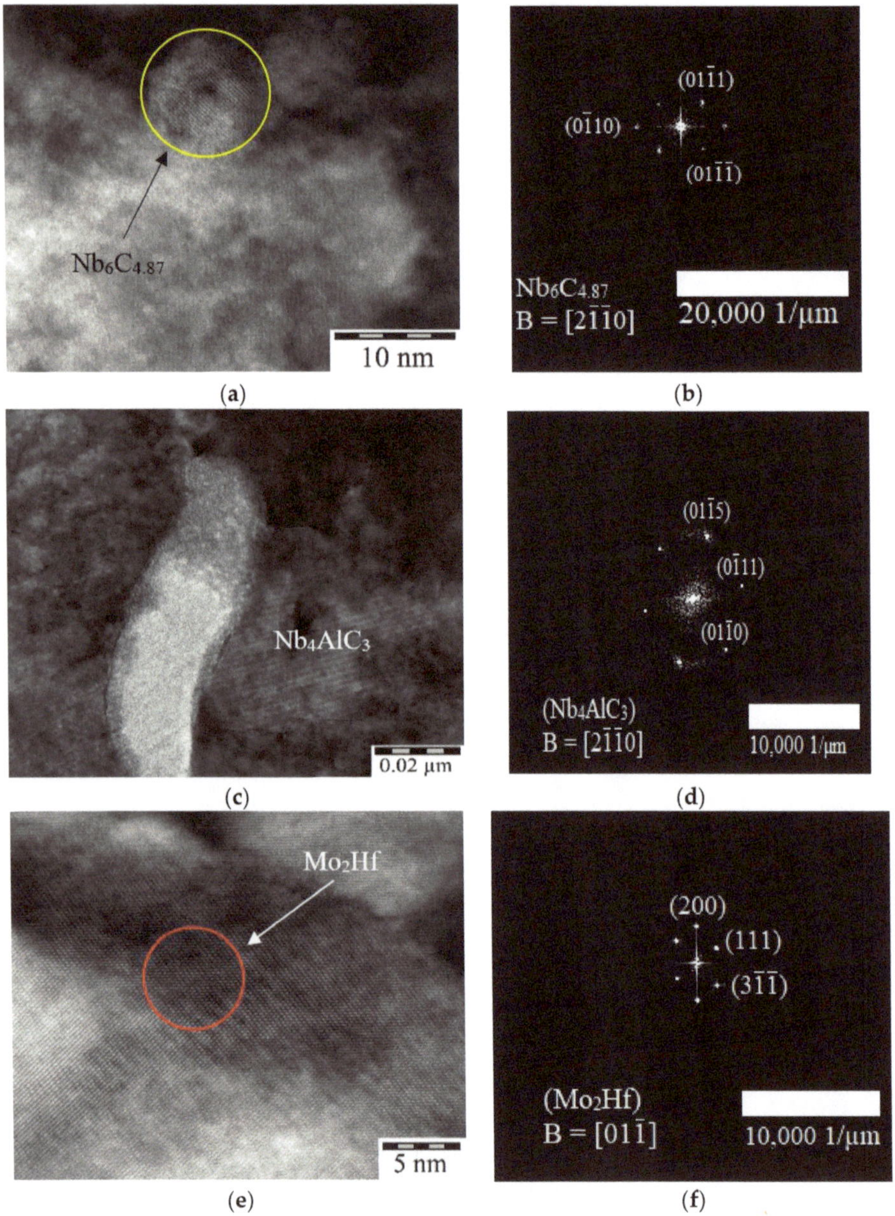

Figure 8. A TEM image of the LPBF samples: (**a**,**c**,**e**) precipitated phases; (**b**,**d**,**f**) Fourier transforms (FFT) recorded for the phase surface.

3.2. The Structural Features of the LPBF Samples after HIP

HIP of the LPBF samples fabricated from the AZhK alloy ensured the formation of a granular microstructure without any characteristic features such as laser tracks or molten pool boundaries (Figure 9a). Furthermore, the volumetric porosity of the LPBF samples decreased to 0.013%.

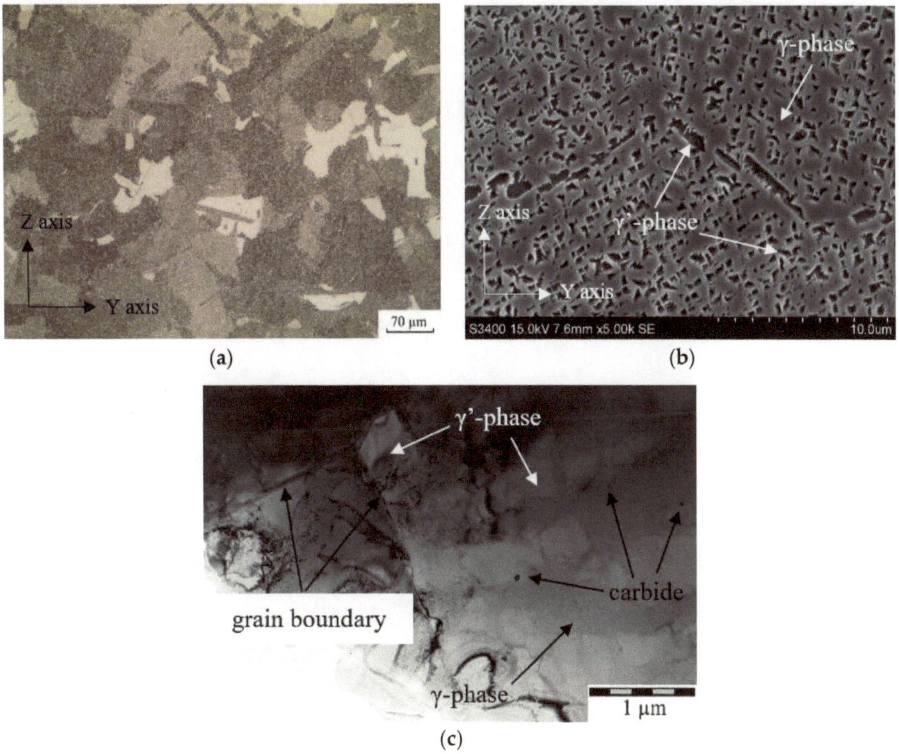

Figure 9. The OM (**a**), SEM (**b**), and TEM (**c**) images showing the microstructure of the LPBF samples after HIP.

Figure 9b shows SEM images of the microstructure of the LPBF samples after HIP. The microstructure consisted of the γ-phase strengthened with Ni_3Al-based intermetallic inclusions (the γ′-phase). The grain boundary was formed of the chains of elongated γ′-phase particles sized up to 3 μm. The γ′-phase was distributed over the bulk of a grain as cubic inclusions up to 1 μm in size, which formed agglomerates consisting of 2–4 particles each (octocubes). According to their morphology, the agglomerates could be classified as the transitional shape between octocubes and octodendrites. The morphology of the γ′-phase is known to predominantly depend on dopants in the alloy and the cooling rate [43]. In this case, the shape and the larger size of γ′-phase grains were related to the low cooling rate after HIP. A TEM image is shown in Figure 9c. After HIP, aggregation of dislocations was no longer observed in the bulk of the sample, which is indicative of internal stress relaxation in the alloy. Nanoparticles with a faceted and spherical shape were detected inside the matrix grains (Figure 10). To determine the composition of the particles, a local EDX analysis was conducted, and the results are presented in Table 5. From the presented results, it can be noted that, in particle 1 (Figure 10a), there is an increased content of Cr, Co, and Mo, and in particle 2 (Figure 10b), there is an increased content of Hf and Al. The obtained results suggest that carbide compounds are formed during the HIP process, which include several elements. An analysis of the FFT images obtained by processing high-resolution images allowed us to infer that they belonged to MeC-type carbides (Ni,Mo)C and (Hf,Al)C.

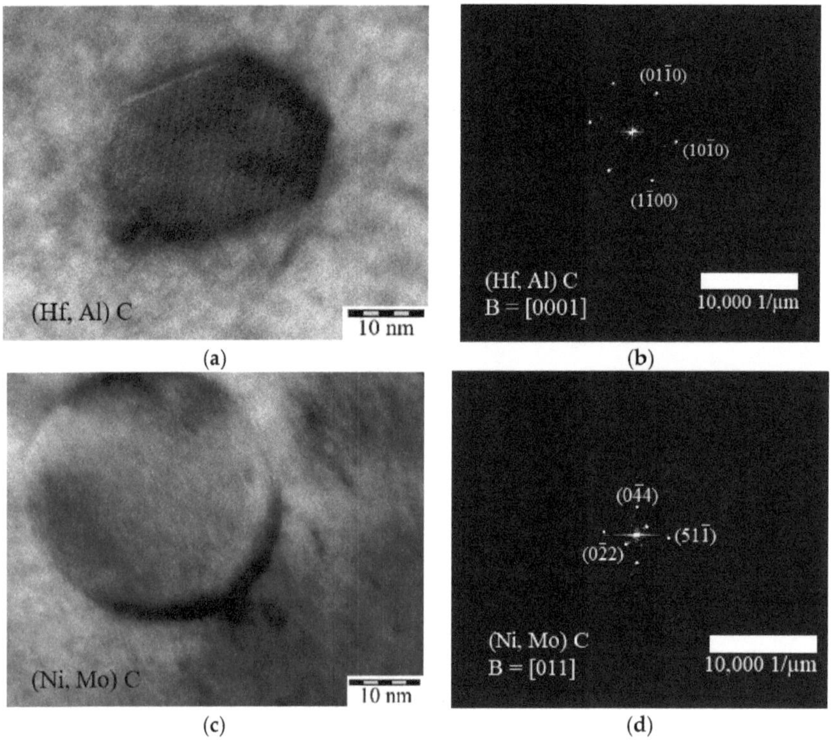

Figure 10. A TEM image of the LPBF samples after HIP: (**a**,**c**) carbides; (**b**,**d**) Fourier transforms (FFT) recorded for the carbide surface.

Table 5. The results of the local EDX analysis of phases in the LPBF sample after HIP.

Spectrum	Content of Elements, wt. %						
	Ni	Cr	Mo	Co	Al	Nb	Hf
1	41.0	20.7	24.3	10.7	3.3	-	-
2	35.8	3.9	-	-	5.8	-	54.5

The lattice parameters of the (Ni,Mo)C carbide were determined by analyzing the FFT image according to interplanar distances between the (0110), (1010), and (1100) planes: a = 11.0262 Å (the tabular lattice parameter was 10.8932 Å). The lattice parameters of the (Hf,Al)C particle with an hcp structure were a = 3.21 Å and c = 28.45 Å (the tabular values were a = 3.32 Å and a = 27.42 Å) (Figure 10c). The lattice parameters of carbides deviated from the tabular values because of the dissolution of nickel or aluminum in them.

3.3. The Structural Features of the LPBF Samples after HIP and HT

After HT, the grain structure was characterized by more well-defined grain boundaries (Figure 11a). The grains were equiaxed; their size in the direction of the Z axis was ≤250 μm. Figure 11b shows the microstructure after HT. One can see that HT contributed to the comminution of the γ'-phase with a cubic morphology. Carbide phases, presumably $M_{23}C_6$-type ones, typical of heat-resistant nickel alloys, were formed at the grain boundaries.

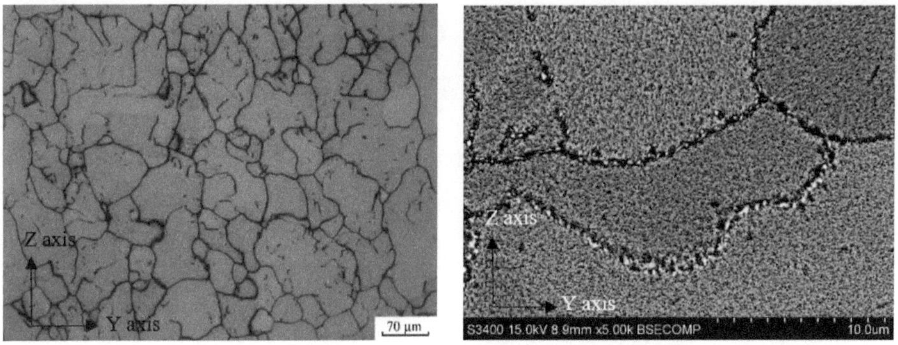

(**a**) (**b**)

Figure 11. The optical microscopy (**a**) and SEM (**b**) images of the structure of the LPBF samples after HIP and HT.

To determine the composition of the precipitated phases at the grain boundary, a local EDX analysis was conducted at points corresponding to the γ/γ'-phases (spectra 6, 8, 10) and directly to the grain boundary itself (spectra 1–5, 7, 9), as shown in Figure 12. The results of the local EMF are summarized in Table 6. The spectra corresponding to the γ/γ'-phases show an increased content of Ni, Cr, Co, and Al, which is typical for these phases. The remaining spectra showed high content of Mo, Cr, or Ni in the particles found at the boundaries and inside the grains. The composition of these particles indicates the formation of carbides of the $M_{23}C_6$ type (Figure 12, Table 6).

Table 6. The results of the local EDX analysis of the composition of the structural components of the LPBF samples after HIP and HT.

Spectrum	Content of Elements, wt. %						
	Ni	Cr	Mo	Co	Al	Nb	Hf
1	18.0	14.9	59.9	7.2	-	-	-
2	8.3	66.5	25.2	-	-	-	-
3	22.2	19.5	54.0	4.3	-	-	-
4	26.9	26.4	34.9	11.8	-	-	-
5	30.5	42.5	21.6	5.4	-	-	-
6	65.7	14.3	5.6	7.8	4.0	2.6	-
7	23.7	27.7	39.5	9.1	-	-	-
8	64.5	14.3	6.1	3.9	6.0	5.2	-
9	8.6	64.2	27.2	-	-	-	-
10	67.5	18.9	4.2	1.9	4.6	2.9	-

Figure 13a shows a TEM image of the grain boundary in the LPBF sample fabricated from the AZhK alloy after HIP and HT. The grain size of the γ'-phase varied within the range from 100 to 250 nm. Figure 11b displays particles that have precipitated along the grain boundaries, which were identified by performing the Fourier transform of the respective zones in the HR-TEM images with the atomic-resolution crystal structure. An analysis of the FFT images (Figure 13d,e) revealed that they belonged to the $Cr_{23}C_6$-type carbide. The calculated lattice parameter for the particle shown in Figure 13b,c was a = 10.92 Å and a = 10.84 Å, respectively, while the tabular value was a = 10.65 Å. The slightly increased (approximately by 2%) unit cell of carbide can be attributed to the dissolution of molybdenum in it.

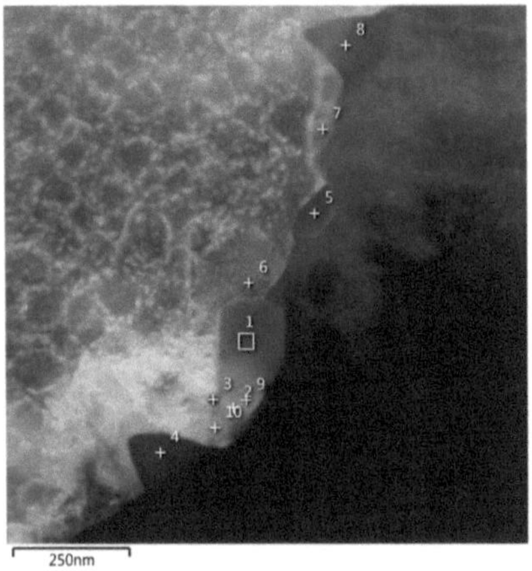

Figure 12. TEM images of the structure of the LPBF samples after HIP and HT, indicating the areas of local EDX.

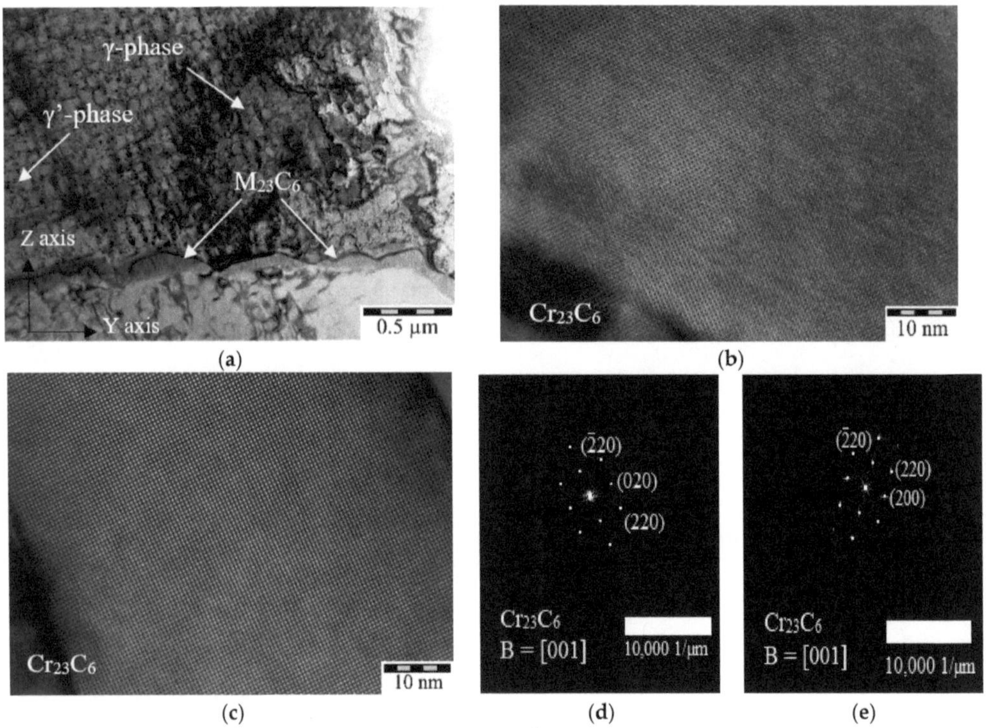

Figure 13. A TEM image of the LPBF samples after HIP and HT. (**a**) The general view; (**b**,**c**) carbides; (**d**,**e**) Fourier transforms (FFT) recorded for the carbide surface.

3.4. Hardness of LPBF Samples after HIP and HT

Table 7 lists the average hardness values for the as-built LPBF samples fabricated from the AZhK alloy and those exposed to HIP and HT in the XY and YZ planes with respect to the build platform. The presented data indicate that hardness of the samples varied negligibly depending on the plane. Hardness in the as-built LPBF samples was low because the nickel matrix contained no strengthening phases. After the HIP procedure, hardness increased by 35 HV units due to precipitation of the γ' phase. The maximum hardness (445 HV) was attained after HT because of uniform segregation of the coherent fine-grained γ'-phase from the nickel matrix. We would like to mention that the AZhK alloy after HT was characterized by high hardness compared to that of other known superalloys, such as Inconel 718 and Inconel 625 [38,44–46].

Table 7. Hardness of LPBF samples fabricated from the AZhK alloy, including after HIP and HT.

Condition of the Alloys	Hardness in the XY Plane, HV	Hardness in the YZ Plane, HV
LPBF	335 ± 30	330 ± 30
LPBF + HIP	370 ± 25	365 ± 20
LPBF + HT	440 ± 30	435 ± 20
LPBF + HIP + HT	445 ± 15	435 ± 15

3.5. Tensile Strength of the LPBF Samples after HIP and HT

The results of room temperature tensile testing of the LPBF samples after various combinations of treatments are summarized in the stress–strain diagram (Figure 14) and in Table 8.

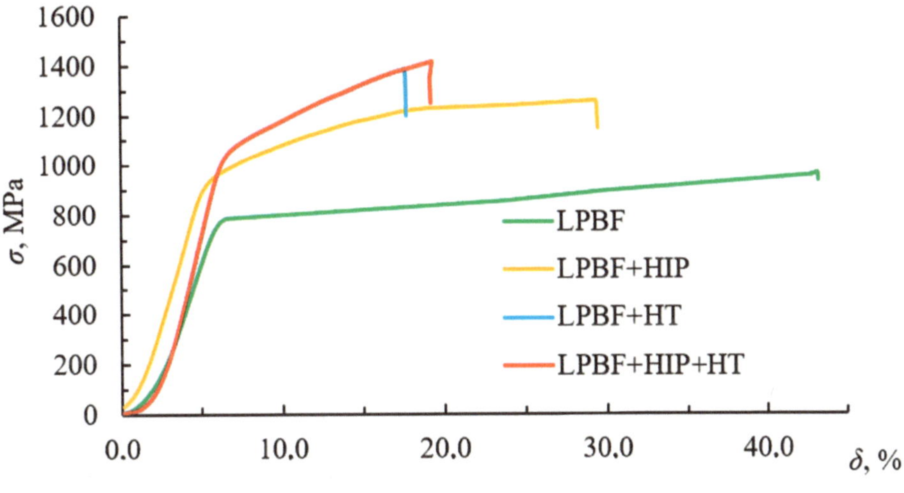

Figure 14. The room temperature stress–strain diagram of the LPBF samples after various treatments.

Table 8. Mechanical properties of the LPBF samples fabricated from the AZhK alloy after HIP and HT at T_{test} = 20 °C.

Condition of the Samples	σ, MPa	$\sigma_{0.2}$, MPa	δ, %
LPBF	970 ± 10	790 ± 15	43.3 ± 3.0
LPBF + HIP	1260 ± 5	860 ± 6	29.4 ± 3.1
LPBF + HT	1395 ± 24	1040 ± 52	17.6 ± 3.0
LPBF + HIP + HT	1410 ± 22	1065 ± 11	19.0 ± 3.0
P/M by HIP	1312	878	14

The relatively low strength and high ductility of this alloy are caused by the fact that the LPBF samples fabricated from the AZhK alloy contained neither strengthening γ'-phase precipitates nor high-melting dopants that would strengthen the γ-phase. It follows from the reported data that post-treatment substantially enhanced the strength of the LPBF samples from 30% (after HIP) to 45% (after HT), while their ductility remained fairly good. Improvement of properties was primarily related to the segregation of the γ'-phase from the supersaturated γ-matrix. The morphology, size, and homogeneous distribution of the coherent γ'-phase in the matrix phase, which ensured the maximum strength of the LPBF samples, played a crucial role in the case of HT. Since the elaborated LPBF mode for the AZhK alloy gave rise to a structure with the fewest defects, the properties of the LPBF + HT and LPBF + HIP + HT samples were very close. The properties of the samples fabricated from the AZhK alloy using the conventional powder metallurgy technology (P/M by HIP) are listed in Table 8 for comparison. The reported data indicate that the required level of mechanical properties of the synthesized AZhK alloy was attained after post-treatment in the HIP + HT or HT modes. Therefore, HT as the manufacturing process was mandatory for fabricating critical parts from the AZhK alloy synthesized by LPBF.

The stress–strain curves and mechanical properties during tensile tests of the synthesized material of AZhK grade at test temperatures of 650 and 750 °C are shown in Figure 15 and Table 9. The tests were performed for the as-built LPBF samples and after HIP + HT.

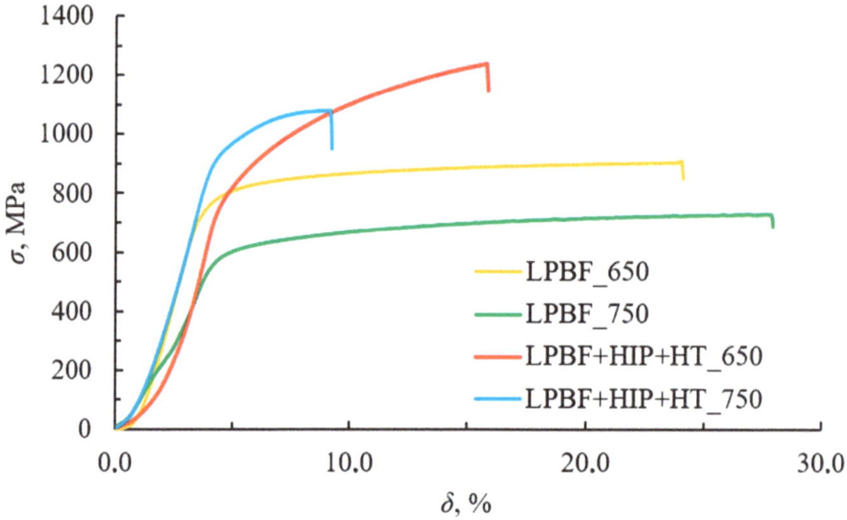

Figure 15. The stress–strain diagram at test temperatures of 650 and 750 °C for the samples after LPBF and LPBF + HIP + HT.

Table 9. Mechanical properties of the AZhK samples in the as-built LPBF and LPBF + HIP + HT conditions at T_{test} = 650 °C and 750 °C.

Condition of the Samples	T_{test}, °C	σ, MPa	$\sigma_{0.2}$, MPa	δ, %
LPBF	650	908 ± 19	695 ± 12	24.0 ± 3.1
	750	718 ± 19	553 ± 29	27.8 ± 2.6
LPBF + HIP + HT	650	1240 ± 25	915 ± 12	15.8 ± 1.5
	750	1085 ± 23	910 ± 24	9.1 ± 2.3
P/M by HIP	650	1202	809	9.5
	750	954	742	17

The ultimate tensile strength and the offset yield point of the LPBF samples are significantly inferior to the nominal values, thus limiting the use of the as-built material. As already mentioned above, the low mechanical properties of the LPBF samples are related to the absence of coherent precipitates of the strengthening γ'-phase and complex carbides in the structure of the material. For the desired level of service properties to be attained, the combined post-treatment in the HIP + HT mode needs to be performed.

Therefore, the structure evolution improved properties of the LPBF samples at 650 °C by: σ, 332 MPa (37%); $\sigma_{0.2}$, 220 MPa (32%); σ, 367 MPa (51%); and $\sigma_{0.2}$, 357 MPa (64%) at 750 °C, respectively. Note that the mechanical properties of the LPBF samples fabricated from the AZhK alloy after HIP + HT were better than those for the samples obtained using the P/M by HIP technology.

3.6. Fabrication of the Ejector-Type Detail from the AZhK Alloy

In accordance with the developed LPBF modes, an ejector-type part was fabricated for the high-melting nickel melt AZhK; its appearance is shown in Figure 3c. The product was intended for reducing the temperature of the working gas from 850 to 450 °C. During the operation, the working gas heated to 850 °C was fed through the central channel; to reduce its temperature, a similar gas with a temperature of 200 °C was fed through two lateral channels. As a result, both gases were ejected, thus reducing the gas temperature in the central region to 450 °C. Due to this process, the temperature of the energy source material decreased, while its speed characteristics remained unchanged, which is an important aspect of engine operation.

The map of deviations of the geometric size of the ejector-type detail fabricated from the AZhK alloy after LPBF (tolerance of ±0.1 mm with respect to the initial electron model) is shown in Figure 16. The findings demonstrate that the deviation between the geometric parameters of the part and the specified values lay above the permissible limit (shown in blue color). These deviations are not critical and do not disrupt the functionality of the part. According to these findings, we have made deviation adjustments to the original electron model.

Figure 16. The map of superimpositions of the initial electronic map and the measured data for the ejector-type part fabricated from the AZhK alloy (permissible deviation ±0.1 mm).

The computed tomography data (Figure 17) have shown that the LPBF mode used for the AZhK alloy ensures the absence of structure discontinuities in the form of incomplete fusion defects, microcracks, and pores sized ≥20 μm, as well as compliance between the geometric parameters of the inner channels for mixing the gas in the electronic model. Channels of the fabricated part after LPBF contained no sintered powder, thus making it unnecessary to perform additional processes for their vibrational cleaning using additional equipment.

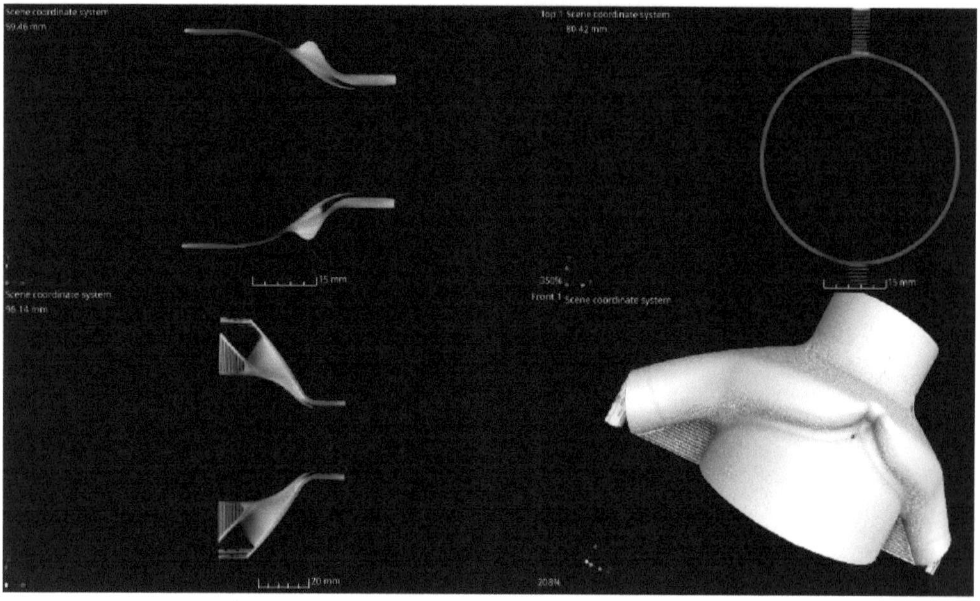

Figure 17. The CT image of the ejector-type detail fabricated from the AZhK alloy.

4. Conclusions

Our conclusions are as follows:

1. The optimal E2 mode of LPBF for the AZhK alloy (E = 125.0 J/mm^3) has been identified, ensuring attainment of a residual porosity of 0.02%. The macrostructure of the LPBF samples consisted of columnar grains oriented in the direction of predominant heat dissipation, perpendicular to the build plane. At the microlevel, the structure was represented by colonies of columnar dendrites. The Nb_4AlC_3 and Nb_6C_4 carbide phases as well as the Mo_2Hf Laves phase were precipitated in the interdendritic region as a result of dopant segregation. The low strength of the LPBF samples manufactured from the AZhK alloy (σ = 967 ± 10 MPa) was related to the absence of strengthening γ'-phase particles and stable carbides.
2. The γ'-phase particles sized up to 2 μm and MC-type Mo- and Hf-based carbides were precipitated during HIP. No Laves phases were detected in the structure of LPBF samples after HIP, as they were dissolved as a result of diffusion. Heat treatment ensured the homogeneous segregation of the fine-grained γ'-phase with a cubic morphology (sized up to 250 nm) and additional precipitation of $Cr_{23}C_6$-type carbides at grain boundaries.
3. The strength and ductility of the LPBF samples fabricated from the AZhK alloy were increased due to solid-solution precipitation of the fine-grained γ'-phase and MC-type carbide phases inside the matrix phase grains and $M_{23}C_6$ at grain boundaries. The following maximum values were obtained: at 20 °C, σ^{20} = 1396 ± 22 MPa and δ = 19.0 ± 3.0 %; at 650 °C, σ^{650} = 1240 ± 25 MPa and δ = 15.8 ± 1.5%; and at 750 °C, σ^{750} = 1085 ± 23 MPa and δ = 9.1 ± 2.3%.
4. A geometrically complex ejector-type part was manufactured using the developed LPBF technology; its geometric parameters corresponded to those of the electronic models, while the base material was free of defects (microcracks or pores) sized more than 20 μm.

Author Contributions: Conceptualization, Z.A.S.; Methodology, F.A.B. and Z.A.S.; Software, E.A.L.; Validation, F.A.B. and Z.A.S.; Formal analysis, Z.A.S.; Investigation, F.A.B., P.A.L., M.Y.B. and I.A.L.; Writing—original draft, F.A.B. and I.A.L.; Supervision, E.A.L. All authors have read and agreed to the published version of the manuscript.

Funding: This work was supported by the Russian Science Foundation (project No. 19-79-10226).

Conflicts of Interest: The authors declare no conflict of interest.

References

1. Armstrong, M.; Mehrabi, H.; Naveed, N. An overview of modern metal additive manufacturing technology. *J. Manuf. Process.* **2022**, *84*, 1001–1029. [CrossRef]
2. Vafadar, A.; Guzzomi, F.; Rassau, A.; Hayward, K. Advances in Metal Additive Manufacturing: A Review of Common Processes, Industrial Applications, and Current Challenges. *Appl. Sci.* **2021**, *11*, 1213. [CrossRef]
3. Cooke, S.; Ahmadi, K.; Willerth, S.; Herring, R. Metal additive manufacturing: Technology, metallurgy and modelling. *J. Manuf. Process.* **2020**, *57*, 978–1003. [CrossRef]
4. Aboulkhair, N.T.; Bosio, F.; Gilani, N.; Phutela, C.; Hague, R.J.; Tuck, C.J. Additive manufacturing processes for metals. In *Quality Analysis of Additively Manufactured Metals*; Elsevier: Amsterdam, The Netherlands, 2023; pp. 201–258. [CrossRef]
5. Katz-Demyanetz, A.; Popov, V.V.; Kovalevsky, A.; Safranchik, D.; Koptyug, A. Powder-bed additive manufacturing for aerospace application: Techniques, metallic and metal/ceramic composite materials and trends. *Manuf. Rev.* **2019**, *6*, 5. [CrossRef]
6. Salunkhe, S.; Rajamani, D. Current trends of metal additive manufacturing in the defense, automobile, and aerospace industries. In *Woodhead Publishing Reviews: Mechanical Engineering Series—Advances in Metal Additive Manufacturing*; Woodhead Publishing: Sawston, UK, 2023; pp. 147–160. [CrossRef]
7. Altıparmak, S.C.; Xiao, B. A market assessment of additive manufacturing potential for the aerospace industry. *J. Manuf. Process.* **2021**, *68 Pt A*, 728–738. [CrossRef]
8. Blakey-Milner, B.; Gradl, P.; Snedden, G.; Brooks, M.; Pitot, J.; Lopez, E.; Leary, M.; Berto, F.; du Plessis, A. Metal additive manufacturing in aerospace: A review. *Mater. Des.* **2021**, *209*, 110008. [CrossRef]
9. Boyer, R.; Cotton, J.; Mohaghegh, M.; Schafrik, R. Materials considerations for aerospace applications. *MRS Bull.* **2015**, *40*, 1055–1066. [CrossRef]
10. Bhat, B.N. (Ed.) *Aerospace Materials and Applications*; American Institute of Aeronautics and Astronautics, Inc.: Reston, VA, USA, 2018. [CrossRef]
11. Akhtar, W.; Sun, J.; Sun, P.; Chen, W.; Saleem, Z. Tool wear mechanisms in the machining of Nickel based super-alloys: A review. *Front. Mech. Eng.* **2014**, *9*, 106–119. [CrossRef]
12. Ezugwu, E.; Bonney, J.; Yamane, Y. An overview of the machinability of aeroengine alloys. *J. Mater. Process. Technol.* **2003**, *134*, 233–253. [CrossRef]
13. Frazier, W.E. Metal Additive Manufacturing: A Review. *J. Mater. Eng. Perform.* **2014**, *23*, 1917–1928. [CrossRef]
14. Froes, F.; Boyer, R. *Additive Manufacturing for the Aerospace Industry*; Elsevier: Amsterdam, The Netherlands, 2019; 482p.
15. Hosseini, E.; Popovich, V. A review of mechanical properties of additively manufactured Inconel 718. *Addit. Manuf.* **2019**, *30*, 100877. [CrossRef]
16. Sanchez, S.; Smith, P.; Xu, Z.; Gaspard, G.; Hyde, C.J.; Wits, W.W.; Ashcroft, I.A.; Chen, H.; Clare, A.T. Powder Bed Fusion of nickel-based superalloys: A review. *Int. J. Mach. Tools Manuf.* **2021**, *165*, 103729. [CrossRef]
17. Volpato, G.M.; Tetzlaff, U.; Fredel, M.C. A comprehensive literature review on laser powder bed fusion of Inconel superalloys. *Addit. Manuf.* **2022**, *55*, 102871. [CrossRef]
18. Han, Q.; Gu, Y.; Setchi, R.; Lacan, F.; Johnston, R.; Evans, S.L.; Yang, S. Additive manufacturing of high-strength crack-free Ni-based Hastelloy X superalloy. *Addit. Manuf.* **2019**, *30*, 100919. [CrossRef]
19. Han, Q.; Mertens, R.; Montero-Sistiaga, M.L.; Yang, S.; Setchi, R.; Vanmeensel, K.; Van Hooreweder, B.; Evans, S.L.; Fan, H. Laser powder bed fusion of Hastelloy X: Effects of hot isostatic pressing and the hot cracking mechanism. *Mater. Sci. Eng. A* **2018**, *732*, 228–239. [CrossRef]
20. Boswell, J.H.; Clark, D.; Li, W.; Attallah, M.M. Cracking during thermal post-processing of laser powder bed fabricated CM247LC Ni-superalloy. *Mater. Des.* **2019**, *174*, 107793. [CrossRef]
21. Wang, X.; Carter, L.N.; Pang, B.; Attallah, M.M.; Loretto, M.H. Microstructure and yield strength of SLM-fabricated CM247LC Ni-Superalloy. *Acta Mater.* **2017**, *128*, 87–95. [CrossRef]
22. Sentyurina, Z.; Baskov, F.; Loginov, P.; Kaplanskii, Y.; Mishukov, A.; Logachev, I.; Bychkova, M.; Levashov, E.; Logacheva, A. The effect of hot isostatic pressing and heat treatment on the microstructure and properties of EP741NP nickel alloy manufactured by laser powder bed fusion. *Addit. Manuf.* **2020**, *37*, 101629. [CrossRef]
23. Baskov, F.; Sentyurina, Z.; Kaplanskii, Y.; Logachev, I.; Semerich, A.; Levashov, E. The influence of post heat treatments on the evolution of microstructure and mechanical properties of EP741NP nickel alloy produced by laser powder bed fusion. *Mater. Sci. Eng. A* **2021**, *817*, 141340. [CrossRef]
24. Raisson, G.; Guédou, J.Y.; Guichard, D.; Rongvaux, J.M. Production of Net-Shape Static Parts by Direct HIPing of Nickel Base Superalloy Prealloyed Powders. *Adv. Mater. Res.* **2011**, *278*, 277–282. [CrossRef]

25. Samarov, V.; Barre, C.; Khomyakov, E.; Haykin, R. Net Shape HIP for Complex Shape PM Parts as A Cost-Efficient Industrial Technology. In Proceedings of the 8th International Conference on Hot Isostatic Pressing, Paris, France; 2005; pp. 48–52.
26. Baccino, R.; Moret, F.; Pellerin, F.; Guichard, D.; Raisson, G. High performance and high complexity net shape parts for gas turbines: The ISOPREC® powder metallurgy process. *Mater. Des.* **2000**, *21*, 345–350. [CrossRef]
27. Bassini, E.; Vola, V.; Lorusso, M.; Ghisleni, R.; Lombardi, M.; Biamino, S.; Ugues, D.; Vallillo, G.; Picqué, B. Net shape HIPping of Ni-superalloy: Study of the interface between the capsule and the alloy. *Mater. Sci. Eng. A* **2017**, *695*, 55–65. [CrossRef]
28. Lacroix, R.; Seifert, R.W.; Timonina-Farkas, A. Benefiting from additive manufacturing for mass customization across the product life cycle. *Oper. Res. Perspect.* **2021**, *8*, 100201. [CrossRef]
29. Attaran, M. The rise of 3-D printing: The advantages of additive manufacturing over traditional manufacturing. *Bus. Horiz.* **2017**, *60*, 677–688. [CrossRef]
30. Westerweel, B.; Basten, R.J.; van Houtum, G.-J. Traditional or Additive Manufacturing? Assessing Component Design Options through Lifecycle Cost Analysis. *Eur. J. Oper. Res.* **2018**, *270*, 570–585. [CrossRef]
31. Kamal, M.; Rizza, G. Design for metal additive manufacturing for aerospace applications. In *Additive Manufacturing for the Aerospace Industry*; Elsevier: Amsterdam, The Netherlands, 2019; pp. 67–86. [CrossRef]
32. Barz, A.; Buer, T.; Haasis, H.-D. A Study on the Effects of Additive Manufacturing on the Structure of Supply Networks. *IFAC-PapersOnLine* **2016**, *49*, 72–77. [CrossRef]
33. Stegman, B.; Shang, A.; Hoppenrath, L.; Raj, A.; Abdel-Khalik, H.; Sutherland, J.; Schick, D.; Morgan, V.; Jackson, K.; Zhang, X. Volumetric energy density impact on mechanical properties of additively manufactured 718 Ni alloy. *Mater. Sci. Eng. A* **2022**, *854*, 143699. [CrossRef]
34. Caiazzo, F.; Alfieri, V.; Casalino, G. On the Relevance of Volumetric Energy Density in the Investigation of Inconel 718 Laser Powder Bed Fusion. *Materials* **2020**, *13*, 538. [CrossRef]
35. Watring, D.S.; Benzing, J.T.; Hrabe, N.; Spear, A.D. Effects of laser-energy density and build orientation on the structure–property relationships in as-built Inconel 718 manufactured by laser powder bed fusion. *Addit. Manuf.* **2020**, *36*, 101425. [CrossRef]
36. Chlebus, E.; Gruber, K.; Kuźnicka, B.; Kurzac, J.; Kurzynowski, T. Effect of heat treatment on the microstructure and mechanical properties of Inconel 718 processed by selective laser melting. *Mater. Sci. Eng. A* **2015**, *639*, 647–655. [CrossRef]
37. Sui, S.; Li, Z.; Zhong, C.; Zhang, Q.; Gasser, A.; Chen, J.; Chew, Y.; Bi, G. Laves phase tuning for enhancing high temperature mechanical property improvement in laser directed energy deposited Inconel 718. *Compos. Part B Eng.* **2021**, *215*, 108819. [CrossRef]
38. Tucho, W.M.; Cuvillier, P.; Sjolyst-Kverneland, A.; Hansen, V. Microstructure and hardness studies of Inconel 718 manufactured by selective laser melting before and after solution heat treatment. *Mater. Sci. Eng. A* **2017**, *689*, 220–232. [CrossRef]
39. Parimi, L.L.; Ravi, G.; Clark, D.; Attallah, M.M. Microstructural and texture development in direct laser fabricated IN718. *Mater. Charact.* **2014**, *89*, 102–111. [CrossRef]
40. Bhavsar, R.B.; Collins, A.; Silverman, S. Use of alloy 718 and 725 in oil and gas industry. In *Superalloys 718, 625, 706 and various derivatives*; Lorin, E.A., Ed.; TMS (The Alinernls. Metals: Materials Society): Pittsburgh, PA, USA, 2001; pp. 47–55.
41. Janaki Ram, G.D.; Venugopal Reddy, A.; Prasad Rao, K.; Reddy, G.M. Microstructure and mechanical properties of Inconel 718 electron beam welds. *Mater. Sci. Technol.* **2005**, *10*, 1132–1138.
42. Li, X.; Shi, J.; Wang, C.; Cao, G.; Russell, A.; Zhou, Z.; Li, C.; Chen, G. Effect of heat treatment on microstructure evolution of Inconel 718 alloy fabricated by selective laser melting. *J. Alloys Compd.* **2018**, *764*, 639–649. [CrossRef]
43. Grosdidier, T.; Hazotte, A.; Simon, A. Precipitation and dissolution processes in γ/γ' single crystal nickel-based superalloys. *Mater. Sci. Eng. A* **1998**, *256*, 183–196. [CrossRef]
44. Popovich, V.; Borisov, E.; Sufiiarov, V.; Masaylo, D.; Alzina, L. Impact of heat treatment on mechanical behaviour of Inconel 718 processed with tailored microstructure by selective laser melting. *Mater. Des.* **2017**, *131*, 12–22. [CrossRef]
45. Feng, K.-Y.; Liu, P.; Li, H.-X.; Sun, S.-Y.; Xu, S.-B.; Li, J.-N. Microstructure and phase transformation on the surface of Inconel 718 alloys fabricated by SLM under 1050°C solid solution + double ageing. *Vacuum* **2017**, *145*, 112–115. [CrossRef]
46. Pleass, C.; Jothi, S. Influence of powder characteristics and additive manufacturing process parameters on the microstructure and mechanical behaviour of Inconel 625 fabricated by Selective Laser Melting. *Addit. Manuf.* **2018**, *24*, 419–431. [CrossRef]

MDPI AG
Grosspeteranlage 5
4052 Basel
Switzerland
Tel.: +41 61 683 77 34

Metals Editorial Office
E-mail: metals@mdpi.com
www.mdpi.com/journal/metals

www.ingramcontent.com/pod-product-compliance
Lightning Source LLC
Chambersburg PA
CBHW042009260926
51850CB00020B/2941